REGULATION OF ENZYME ACTIVITY

Series editors

David Rickwood
Department of Biology, University of Essex, Wivenhoe Park, Colchester, Essex CO4 3SQ, UK

David Male
Institute of Psychiatry, De Crespigny Park, Denmark Hill, London SE5 8AF, UK

REGULATION OF ENZYME ACTIVITY

J.H.Ottaway

University of Bradford, School of Studies in Environmental Science,
Bradford, West Yorkshire BD7 1DP, UK

OXFORD · WASHINGTON DC

IRL Press
Eynsham
Oxford
England

© IRL Press Limited 1988
First published 1988

British Library Cataloguing in Publication Data

Ottaway, J.H. (James Henry), *1923-*
    Regulation of enzyme activity.
    1. Organisms. Enzymes. Regulation
    I. Title   II. Series
    574.19′25

Library of Congress Cataloging in Publication Data

Ottaway, J. H. (James Henry)
    Regulation of enzyme activity.

    (In focus)
    Includes bibliographies and index.
    1. Enzymes--Regulation. I. Title. II. Series:
In focus (Oxford, England) [DNLM: 1. Enzyme Activation.
2. Enzymes--metabolism.   QU 135 089r]
QP601.5.087 1988   574.19′25   88-13161
ISBN 1 85221 072 9 (soft)

Typeset by Infotype and printed by Information Printing Ltd, Oxford, England.

Preface

This book is not a guide to metabolic regulation, of which many good examples exist; those quoted in Chapter 1, Section 5.2 may be especially recommended. It is, instead, a book on the ways in which enzymes may be, and are, reversibly regulated, by ligand binding, by covalent modification, or by other means. I had hoped to include a description of the ways in which the activity of enzymes may be regulated by changes in their rate of synthesis or degradation. However, the need to provide generous illustration of the text made it impossible, within the planned compass of the book, for even a brief survey of this very important mode of enzyme regulation.

A good deal of trouble has been taken to make the review wide-ranging, by including as much information as possible, within the limits of space imposed by the publishers, about enzyme regulation in plants and bacteria. In order to do this, it has consequently been necessary to omit a good deal of information of the sort usually found in books of this type; for example the details of glycogen metabolism is not described in its entirety because admirable summaries can be found in the books recommended at the end of Chapter 1. On the other hand, the survey is as topical as possible, in a rapidly expanding field, so that many of the references, often to reviews, date from the last few years. It is hoped that readers will find the examples that have been chosen helpful and illuminating.

I would like to dedicate this book to my first—and, sometimes, during the last 15 years, my only—convert to the delights and rigours of metabolic flux analysis, Dr Linda Saunderson (née McMinn). My thanks to my daughter Sabina for typing the manuscript.

J.H.Ottaway

Contents

3. Regulation of ligand binding

4. Regulation by reversible covalent modification

5. Signals, transducers and modifiers

6. Regulation of enzymes in plants

7. Overview

Abbreviations

Ala	alanine
AMP	adenosine 5′ monophosphate
ADP	adenosine 5′ diphosphate
Asp	aspartate
ATP	adenosine 5′ triphosphate
c_{E_i}	flux control coefficient for the ith enzyme in a pathway
CaM	calmodulin
cAMP	cyclic AMP
CAT	carbamoylphospate – aspartate transferase
cAMP – PK	cAMP-dependent protein kinase
cGMP	cyclic GMP
CoA	coenzyme A
CRP	cAMP receptor protein
CTP	cytidine 5′ triphosphate
DAG	diacylglycerol
dATP	deoxyadenosine 5′ triphosphate
dNTP	(unspecified) deoxynucleoside 5′ triphosphate
F-1,6-BP	fructose-1,6-bisphosphate
F-2,6-BP	fructose-2,6-bisphosphate
F-6-P	fructose-6-phosphate
FBPase 1(2)	fructose-1(2),6-bisphosphatase
G-6-P	glucose-6-phosphate
GAPDH	glyceraldehyde-3-phosphate dehydrogenase
Glc	glucose
Gln	glutamine
Glu	glutamate
GS	glutamine synthase
His	histidine
HMG-CoA	hydroxymethylglutaryl-CoA
IDH	isocitrate dehydrogenase
K_{eq}	equilibrium constant
K_m	Michaelis constant
lac	lactose
MDH	malate dehydrogenase
MLCK	myosin light chain kinase

NAD(P)$^+$	nicotinamide-adenine dinucleotide (phosphate)
NAD(P)H	reduced nicotinamide – adenine dinucleotide (phosphate)
NTP	(unspecified) nucleoside triphosphate
αOG	α-oxoglutarate
OAA	oxaloacetate
OADH	oxoacid dehydrogenase
P_i	inorganic phosphate
PP_i	inorganic pyrophosphate
PDE	(cyclic nucleotide) phosphodiesterase
PDRP	pyruvate dikinase regulatory protein
PEP	phosphoenolpyruvate
PFK1(2)	phosphofructokinase producing F-1,6-BP or F-2,6-BP, respectively
PGA	phosphoglyceric acid
PIP_2	phosphatidylinositol bisphosphate
PI_3	inositol triphosphate
PK	protein kinase
PK-C	protein kinase C
Pyr	pyruvate
redox	oxidation/reduction
mRNA	messenger RNA
Ser	serine
$t_{1/2}$	time taken for a reaction to reach 50% completion
TDO	tryptophan-2,3-dioxygenase
THFA	tetrahydrofolic acid
UDP-Gal	uridine diphosphate – galactose
UDP – Glc	uridine diphosphate – glucose
UMP	uridine monophosphate
v	velocity of an (enzyme-catalysed) reaction
V, V_{max}, V_s	value of v when an enzyme is saturated with substrate
Z	sensitivity coefficient

1

Amplification, time scales and feedback

<hr>

> 'Was this the face that launch'd a thousand ships?'
> *Marlowe*

1. Introduction

For reasons of space this book does not deal with irreversible cascades, such as blood clotting or complement formation, which have the nature of a rapid activation followed by a 'damage limitation' exercise. Only reversible systems are considered, and there are basically two types: systems with, or without, amplification. Systems with no amplification are those in which an end-product or allosteric modifier accumulates or disappears, and changes the activity of an enzyme. The response may be non-linear (e.g. a sigmoidal curve of v against [S], Chapter 3) but the *gain* is unity. Systems with amplification typically have a cascade mechanism, and a gain which can be much greater than unity. The properties of such cascades are discussed in the next section.

2. Features of amplification mechanisms

Stadtman and co-workers (1) have investigated the regulatory properties of reversible cyclic cascades, and have classified them under three main headings, in terms of a notional monocyclic cascade (*Figure 1.1*), using reversible phosphorylation of an enzyme as an example.

2.1 Signal amplification

This is given by the following quotient: the concentration of modulator ligand L_1 required to give 50% activation of a converter enzyme C to C_a, divided by the concentration of L_1 required to give 50% of the interconvertible enzyme I in its modified form. Typically this quotient is greater than unity, that is unless

1

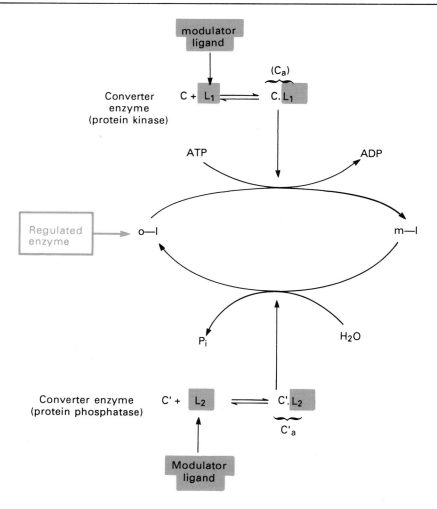

Figure 1.1. An open monocyclic reversible enzyme cascade. When the regulated enzyme has been modified (m-I), it may be either in the activated or inactivated form.

the re-conversion of modified I (m-I) back to its original form (o-I) is very rapid; 50% conversion of the total I present to m-I can be accomplished with much less than 50% of C in the active (C_a) form. The ratio increases exponentially with the number of cascades. In practice, the concentration of L_1 can be very small indeed. How big the amplification is will depend on the speed of the dephosphorylation, and this will depend on the concentration of L_2, among other factors. The numerical value of signal amplification depends on the parameter values, but a computer simulation with arbitrary parameter values gave values of $\approx 300^n$, where n is the number of stages in the cascade. In practice, the value may be significant, but a good deal less than this; for glutamine synthase, for

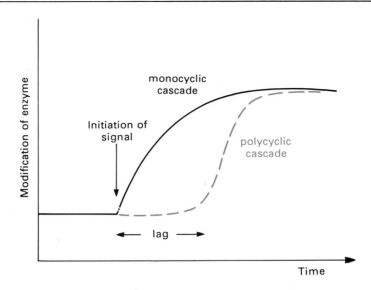

Figure 1.2. Time relationships for enzyme modification by monocyclic and polycyclic cascade systems.

example, a value of 67 was found experimentally for a single stage, and 250–1500, depending on conditions, for a bicyclic cascade (2).

The catalytic amplification, which is simply put as

$$[\text{m-I}]_{max}/[C_a]_{max} \qquad [1.1]$$

is a quite different ratio, and not of great interest. ([m-I] can be replaced by [o-I], if the latter is the active form of I.)

2.2 Amplitude

This is the maximum (fractional) value for the activation of I that is possible, if the concentration of L_1 [L_1] is raised to a saturating level. The maximum value, perhaps surprisingly, need not be unity, as for glycogen synthase, which has several sites that can be phosphorylated, perhaps by different kinases, but is in any event a smooth function of [L_1], not a step function (trigger mechanism).

2.3 Changes of rate

For a monocyclic cascade (*Figure 1.2*), the rate of conversion from one steady state to another is governed (3) by an exponential function of the form

$$I = I_{new}(1 - e^{-kt}) \qquad [1.2]$$

(where k is a rate constant).

For a polycyclic cascade there is always a lag period as the amplifying enzymes become activated (exaggerated in *Figure 1.2*), followed by a change which can be very rapid. For example, the half reaction time for conversion of phosphorylase *b* to phosphorylase *a* in frog sartorius muscle is estimated at 700 ms.

2.4 Sensitivity

This is used in a different sense than in Chapter 2, because it arises from a different cause. If an allosteric ligand—for example α-oxoglutarate in *Figure 4.7b*—has a positive effect on more than one step in the cascade, the fractional modification versus [ligand] curve will be sigmoidal, whereas if a ligand—for example, glutamine in *Figure 4.7b*—has a negative effect, the effector curve will be 'over-square'.These changes do not arise because of cooperativity between multiple binding sites on the same protein, but because these variables appear in more than one of the equations describing the cascade. The definition is not reproduced here, but is so arranged that a sigmoidal curve will give a maximum 'sensitivity' of 3.3 for a monocyclic cascade and 7.9 and 12.5 for bi- and tricyclic cascades, with values of less than one for antagonistic effectors.

Readers will no doubt be familiar with multicyclic cascades—the adrenergic activation of glycogen phosphorylase, for example, has four cycles. The analysis of expected behaviour becomes correspondingly complex (1), but experimental verification, using a synthetic mixture of purified components from the glutamine synthase cascade (see Section 2.2 of Chapter 4) has shown very good agreement with predicted behaviour (4). It is worth mentioning in passing that multistage cascades can be closed, in which the activating and inhibiting catalyst reside on the same protein—as is the case in the regulation of glutamine synthase—or open, when the activating catalyst, for example a protein kinase, and the inhibiting catalyst, for example a phosphatase, are different entities. This is more common.

3.   Time scales of enzyme regulatory mechanisms

In a cascade system the duration of an 'on' switch is one means of providing amplification, providing that the messenger which is synthesized during the 'on' period is not rapidly removed. The activation of transducin by activated rhodopsin (see Section 1.1 of Chapter 5) is a very good example of such a phenomenon. A separate consideration is *response time*; that is the time interval before the response reaches a steady state. The prize in the slow bicycle race appears at present to be held by the inactivation of the branched chain oxoacid dehydrogenase complex (*Figure 1.3*), which takes two days to reach completion when measured *in vivo*, and is not well correlated with the disappearance of an 'activator protein' (5). *In vitro*, on the other hand, the half reaction time is at most 25 min (6), which illustrates the difficulties sometimes encountered in translating *in vitro* studies to behaviour in the intact animal. The pyruvate dehydrogenase complex is notoriously slow, taking about 20 – 30 min to move

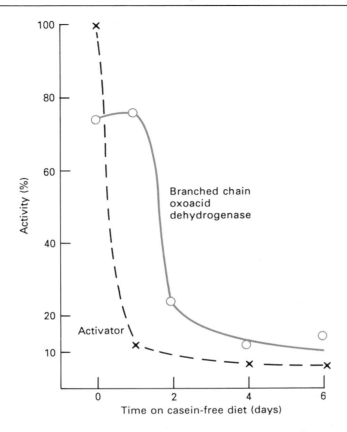

Figure 1.3. Change in activity of branched chain oxoacid dehydrogenase activity in rat-liver mitochondria after the animals have been placed on a low-protein diet.

from complete activation to complete inactivation, or vice versa (7). At the other end of the time sale, activations which may involve covalent modifications are rarely complete in less than a second; the activation of phosphorylase is an example and here we can note the advantage, in metabolic control, of differences in time scale. The activation of glycogen synthase is notably slower than that of phosphorylase (8) (*Figure 1.4*), which makes considerable sense in avoiding futile cycling of glucose residues. There are, of course, even faster responses; the response of a protein to ligand binding can occur in milliseconds, but it may take an elaborate apparatus to procure a reversible change of any magnitude in the concentration of ligand within the time scale (see *Figure 5.5*).

These enormous differences in time scales have to be borne in mind constantly. Not only do they give great flexibility in control, but it does not by any means follow that when two or more enzymes are regulated in a pathway, their response times will be similar. In a metabolic pathway, diagram arrows carrying '+' and '−' signs with no information about response times can be seriously misleading.

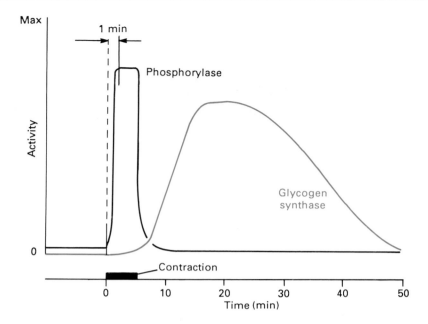

Figure 1.4. Time scale of activation of phosphorylase and glycogen synthase during and after a 5 min contraction of skeletal muscle (derived from ref. 8).

4. Feedback control

Feedback, for electronic and control engineers, is a very serious topic with wide ramifications into such matters as oscillations (9), stability, and proportional gain. Here we shall avoid such advanced topics and discuss only positive and negative feedback, and positive feedforward mechanisms.

It is not essential for a regulatory enzyme to be controlled by a feedback mechanism at all. Hormone-sensitive triglyceride lipase of adipose tissue cells, for example, does not have any end-product control, and it is possible for very steep rises in plasma non-esterified fatty acid levels to occur as a result. There is a recovery system within the fat cells which resynthesizes triacylglycerols from monoacylglycerol and free fatty acids, but this is in no way a feedback.

The point to remember is that if feedback control exists, it must act in such a way as to bring the flux through the enzyme back to a 'normal' level. In the ordinary way, one thinks in this context of an accumulation of end-products which allosterically inhibit the activity of an enzyme near the beginning of a metabolic pathway; a very complex example is the regulation in bacteria of the synthesis of amino acids from aspartate (10). However, the inhibition of pyruvate dehydrogenase kinase by coenzyme A (CoA) (11), which has the effect of increasing the rate of production of acetyl units, is an example of a release of inhibition of flux by a product; it is one of the products of the disposal of acetyl-

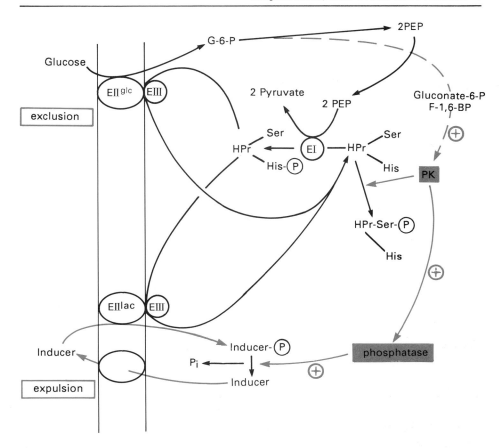

Figure 1.5. Mechanism of sugar uptake in Gram-positive bacteria. The upper half of the figure focuses on the positive feedback inherent in the mechanism. Enzyme EI catalyses the transfer of a phosphoryl group from PEP to the small carrier protein, HPr. EIII catalyses transfer of the phosphoryl residue to one of a series of integral membrane permeases (EII), of which two are shown. These in turn phosphorylate the external sugar molecule, coupling this with transfer through the membrane. The lower half of the figure, and the parts printed in orange, show the regulation of the permease system both by inactivation of HPr by a protein kinase (PK), and by expulsion of inducers of new permease synthesis such as thio-β-methyl galactoside. Ser, serine; His, histidine; Lac, lactose; Glc, glucose.

CoA generated by pyruvate oxidation (citrate being the other) and could be classed as negative feedback.

Positive feedback does exist, but it has to be very tightly regulated if control of a system is to be maintained. For example glucose transport in Gram-positive bacteria is by means of a phosphate-exchange enzyme (12) combined with a membrane-bound transporter (*Figure 1.5*). Since each molecule of glucose-6-phosphate (G-6-P) can give rise to two molecules of phosphoenolpyruvate (PEP), this system is inherently unstable, and it is controlled by inactivation of an essential component, the phosphocarrier protein HPr, by a protein kinase activated by intermediates of glycolysis or the pentose phosphate pathway (12). A description of details of positive feedback leading to oscillations is given elsewhere (9). Positive *feedforward* is a phenomenon which is not very usual in control engineering, because it can lead to unstable situations, but it is not uncommon in biochemical systems. In its simplest form, it consists of the activation of an enzyme by its substrate, either directly by homotropic cooperativity (Chapter 3), or indirectly through the action of a kinase. For example coenzyme A is a substrate of pyruvate dehydrogenase, and the inhibition of the kinase described above could be described equally well as positive feedforward or negative feedback. Pyruvate, the main substrate of the complex, also inhibits the kinase, giving a more clear-cut example of positive feedforward. A more sophisticated example, also taken from carbohydrate metabolism, is the allosteric activation of yeast and liver pyruvate kinase (13,14) by fructose-1,6-bisphosphate (F-1,6-BP). The effect of this is that an increase in glycolytic flux, with an accompanying increase in activity of phosphofructokinase, and increase of concentration of F-1,6-BP will lead to activation of the terminal enzyme of the sequence *before* the 'slug' of metabolites reaches it. Since pyruvate (or acetaldehyde) is the major acceptor of reducing equivalents in anaerobic glycolysis, this ensures that the central part of the glycolytic chain can also work at maximum throughput, limited only by supply of ADP and inorganic phosphate. The system is stable, because as the concentration of F-1,6-BP falls, so pyruvate kinase loses activity. One can also see that such a system would work very well when there is lack of correspondence between the response times of an early and a late enzyme in a sequence.

5.   Further reading

5.1 Further reading for Chapter 1

Riggs,D.S. (1967) *Adv. Enzyme Regul.,* **5**, 357.

Segel,L.A. (1984) *Modeling Dynamic Phenomena in Molecular and Cellular Biology.* Cambridge University Press, Cambridge.

Stadtman,E.R. (1970) In *The Enzymes*, Volume 1. Boyer,P.D. (ed.), Academic Press, New York, 3rd edn. (An invaluable guide to all types of feedback regulation in bacteria and animal cells.)

5.2 Recommended general reading

Cohen,P. (1976) *Control of Enzyme Activity* (Outline Studies in Biology Series). Chapman & Hall, London.

Martin,B.R. (1987) *Metabolic Regulation—a Molecular Approach.* Blackwell Scientific Publishers, Oxford.

Ochs,R.S., Hanson,R.W. and Hall,J. (eds) (1985) *Metabolic Regulation.* Elsevier, Amsterdam.

6. References

1. Stadtman,E.R. and Chock,P.B. (1978) *Curr. Top. Cell. Regul.,* **13**, 53.
2. Mura,U., Chock,P.B. and Stadtman,E.R. (1981) *J. Biol. Chem.,* **256**, 13022.
3. Chock,P.B., Rhee,S.G. and Stadtman,E.R. (1980) *Annu. Rev. Biochem.,* **49**, 813.
4. Chock,P.B., Shacter,E., Jurgensen,S.R. and Rhee,S.G. (1985) *Curr. Top. Cell. Regul.,* **27**, 3.
5. Beggs,M., Espinal,J., Patel,H. and Randle,P.J. (1986) *Biochem. Soc. Trans.,* **14**, 1055.
6. Randle,P.J., Patston,P.A. and Espinal,J. (1987) In *The Enzymes*, Volume 18B. Boyer,P.D. and Krebs,E.G. (eds), Academic Press, New York.
7. Linn,T.C., Pettit,F.H. and Reed,L.J. (1969) *Proc. Natl. Acad. Sci. USA,* **42**, 234.
8. Martin,B.R. (1987) *Metabolic Regulation—A Molecular Approach.* Blackwell Scientific Publishers, Oxford, p. 230.
9. Hess,B. and Markus,M. (1987) *Trends Biochem. Res.,* **12**, 45.
10. Cohen,G.N. (1969) *Curr. Top. Cell. Regul.,* **1**, 183.
11. Reed,L.J., Damuni,Z. and Merryfield,M.L. (1985) *Curr. Top. Cell. Regul.,* **27**, 41.
12. Saier,M.H. (1985) *Mechanisms and Regulation of Carbohydrate Transport in Bacteria.* Academic Press, Florida, p. 12.
13. Tejwani,G.A. (1985) In *Metabolic Regulation.* Ochs,R.S., Hanson,R.W. and Hall,J. (eds), Elsevier, Amsterdam, p. 46.
14. Seubert,W. and Schoner,W. (1971) *Curr. Top. Cell. Regul.,* **3**, 237.

2

Theoretical aspects of the regulation of enzyme activity

A first-rate theory predicts; a second-rate theory forbids; and a third-rate theory explains after the event.
Alex Kitaigorodskii

1. Introduction

In order to see what factors operate in regulation by an enzyme in a linear system, let us take the simplest such reversible system (where S equals substrate and P, product)

$$S \; \overset{E_1}{\rightleftharpoons} P_1$$

$$P_1 \; \overset{E_2}{\rightleftharpoons} P_2$$

for which steady-state rate equations can be written in terms of the concentrations of the two enzymes E_1 and E_2, $[E_1]$ and $[E_2]$ respectively, and the various limiting rates V and K_m values of the enzymes. If the system as a whole is in a steady state, then

$$v_1 = v_2 = v$$

Since we are interested in how the system changes when the activity (i.e. concentration) of one of the enzymes changes, differentiate v with respect to $[E_1]$, say, assuming $[E_2]$ remains constant. We get

$$\frac{\delta v}{\delta [E_1]} = \frac{-(b_1 v + c[E_2])}{2av + b_1[E_1] + b_2[E_2]} \qquad [2.1]$$

10

where a, b and c are complex expressions involving the kinetic constants and S and P_2. A similar operation could be carried out for $[E_2]$.

If we divide the numerator of the differential term by v and the denominator by $[E_1]$, we shall obtain what control engineers call a *sensitivity coefficient*, that is, the measure of the fractional response of the system to a fractional change in a single variable.

$$Z_1 = \frac{\delta v}{v} \cdot \frac{\delta[E_1]}{[E_1]} = \frac{-[E_1](b_1 v + c[E_2])}{2av^2 + (b_1[E_1] + b_2[E_2])v} \qquad [2.2]$$

Z can vary between zero and 1; let us look at three possible values, namely 1, 0 and 0.5. These conditions allow us to simplify Equation 2.2. If Z_1 is equal to one or zero then either $[E_2]$ or $[E_1]$, respectively, disappear from the rearranged equation—results that might be expected. For the remaining enzyme concentrations to have positive (i.e. physically meaningful) values, a complex inequality involving, in each case, $[S]$, $[P_2]$, their Michaelis constants, and V_1 and V_2, has to be satisfied. This may or may not be possible for any pair of enzymes; note that it is not automatically true that E_1 is the 'pacemaking' enzyme (i.e. Z_1 is not automatically 1).

If Z_1 is equal to 0.5, v disappears entirely from Equation 2.2 and we get

$$[E_1]/[E_2] = b_2/b_1 \qquad [2.3]$$

Neither b_1 nor b_2, when referred back to the original rate equations, contains any negative terms, so that it must *always* be possible to find a ratio of $[E_1]/[E_2]$ for which Z_1 is equal to 0.5 (and Z_2 equals 0.5). When Equation 2.3 is satisfied, control is shared equally between the two enzymes. Of course, we cannot say whether this is likely in any 'physiological' situation.

This type of analysis could be extended to any number of enzymes, but it would become impossibly complex, and very few precise conclusions could be drawn from it. It is, however, possible to adopt a similar approach either by computer simulation, a technique pioneered by the author (1,2), or by using results from appropriately designed experiments. A generalized analysis with its practical applications is outlined in the next section.

2. An outline of metabolic control theory

The treatment used here is based on that of Kacser and Burns, originally published in 1973 (3) and described further elsewhere (4). A recent and reliable assessment has been published, see (5) and Sections 5 and 6. The term sensitivity coefficient is used in different senses by different people, and by international agreement, for metabolic control analysis it is now called the flux control coefficient, with the symbolic $c_{E_i}^j$, where j would only have a value greater than one if there existed an interlocking network of pathways. This symbolism is rather clumsy, and in this chapter the superscript j will be omitted. The symbol i ($i = 1$ to n) locates a single step, whether facilitated or catalysed, in the n such steps in the pathway. To define i in this way minimizes the danger of equating the coefficients c_{E_1}, c_{E_2}, etc. with the enzymes E_1, E_2 . . ., while forgetting the boundary flux, which may be very important, although not enzymatic (see Section 5.2).

2.1 Flux summation property
The equation

$$\Sigma c_{E_i} = 1 \qquad [2.4]$$

for a single pathway (i.e. $j = 1$), is the most important prediction of the analysis, because it enables one to assess the relative importance of all the enzymes in a pathway when it is operating in a steady state. There has been some argument about its universal validity, but in my opinion, Equation 2.4 holds exactly for the circumstances for which it was derived; in other situations the approximation

$$\Sigma c_{E_i} \simeq 1 \qquad [2.4a]$$

would be appropriate, but the difference is not as critical as it has been argued to be. For references to these arguments, see Section 5. Three examples of the use of Equation 2.4 are given later in this chapter, and other examples are given elsewhere (5).

2.2 Elasticity or elasticity coefficient
The elasticity or elasticity coefficient relates the local response of one enzyme to a fractional change in one of its substrates, products or effectors

$$\epsilon_{X_i}^{j} = \frac{(\delta v_j / v_j)}{(\delta X_i / X_i)} \qquad [2.5]$$

where v_j is the local flux through enzyme E_j (not necessarily the overall system flux), and X_i is the ith metabolite in the system. This coefficient is related to the flux control by the connectivity property.

2.3 Connectivity property
In its simplest form, this relates the flux control coefficient of adjacent enzymes (E_j, E_{j+1}) to their elasticity coefficients with respect to the shared metabolite, X.

$$c_{E_j} / c_{E_{j+1}} = \epsilon_X^{j+1} / \epsilon_X^{j} \qquad [2.6]$$

In theory, if all the conditions were reproduced, one could measure the elasticities in Equation 2.6 *in vitro*, and then, if c_{E_j} is known, use the relationship to estimate $c_{E_{j+1}}$. In practice, even so basic a parameter as the pH is often not known accurately. In such cases one can use the experimentally measured ratios (Γ, see ref. 5) between the concentrations of neighbouring metabolites, where Γ is given by

$$\Gamma \cdot K_{eq} = \frac{[X_{i+1}]}{[X_i]} \qquad [2.7]$$

to obtain a relationship between c_{E_j} and $c_{E_{j+1}}$. Note that Γ has been used for many years to point to the existence of 'crossover points', that is steps in a pathway where product and substrate are far from equilibrium (K_{eq} is the equilibrium constant) (6). Equation 2.6 can be generalized (5,7) to

$$\sum_{i=1}^{n} c_{E_i} \cdot \epsilon_{X_k}^i = 0 \qquad [2.8]$$

where X_k is any of the m metabolites of the system (but not a completely external effector (which could be Z in the notation described in Section 4.1).

The purpose of using such relationships is to establish a complete set of control coefficients in cases where it is not possible to perturb experimentally every enzyme in a system. Examples will be found in the papers quoted below. Extensions of the original theorems to moiety-conserved cycles (24) (i.e. those in which relationships exist, such as [ATP] + [ADP] = constant), and branched pathways (8,9) have been published, also rules for using matrix algebra to derive the relationships (10), but to describe them is beyond the scope of this book. It is more rewarding to look at a few examples of Equation 2.4 in use.

3. Examples of metabolic control analysis in use

3.1 The triose phosphate segment of glycolysis

This was derived from a computer simulation (11) of aldolase, triose phosphate isomerase, and a simplified equivalent of glyceraldehyde phosphate dehydrogenase (GAPDH). The source was provided by differing, but constant, concentrations of fructose-1,6-bisphosphate (F-1,6-BP), while GAPDH acted as a concentration-dependent transport to the 'ultimate sink'.

Table 2.1 shows that the sum of the flux control coefficients always adds up to 1.0, but that the partition between them can lie completely with aldolase or completely with GAPDH, depending on the concentration of F-1,6-BP and the concentration of aldolase (NB 10^{-4} M is a realistic value for the latter). An important point is that it is legitimate to isolate a section of the pathway for particular study, and Σc_{E_i} will be 1 for the number of enzymes studied, but if this section is then embedded in a more complete pathway, the absolute values of the flux control coefficients will fall, and may be relatively unimportant in

Table 2.1. Flux control coefficients for triose phosphate metabolism in varying conditions

	Low aldolase [1 μM]		High aldolase [100 μM]	
Substrate (F-1,6-BP) concentration (μM)	1	50	1	50
Flux through system (μmol kg^{-1} min^{-1})	0.3	2.3	18	62
	Control coefficients			
Aldolase	0.99	0.99	0.54	0.07
Triose phosphate isomerase	0.003	0.004	0.13	0.02
GAPDH	0.005	0.007	0.33	0.91
Sum	0.998	1.001	1.00	1.00

the partition of values for the extended system. This is true for glycolysis, where phosphofructokinase and hexokinase are much more important overall than aldolase or GAPDH (4).

3.2 Tryptophan catabolism

This study is interesting because it shows experimentally that control may pass from one step to another with changing conditions, and also because it introduces the complications to be found in branched pathways. *Figure 2.1* is a summary of tryptophan catabolism. The flux was measured through the irreversible step catalysed by tryptophan-2,3-dioxygenase (TDO; II in *Figure 2.1*). The activity (amount) of TDO can be increased almost 8-fold, and the catabolic flux of tryptophan almost 4-fold, by the treatment of the animal with the inducing agent dexamethasone. Pyridoxine deficiency, on the other hand, decreases the activity of kynureninase (IV in *Figure 2.1*), for which pyridoxal phosphate is a prosthetic group. For details of the methods used either to estimate a point value for dJ/dE_i (where J is the flux through the system) or to derive c_{E_i} indirectly, the reader is referred to the original papers (12,13). *Table 2.2* summarizes some of the results.

Several interesting points arise from these data. The first is that control never resides completely with TDO, although it breaks the pyrrole ring of tryptophan irreversibly; indeed, when TDO is maximally induced, the major limiting factor is transport of tryptophan into the cell. Secondly, in animals that do not have a vitamin deficiency, these two control coefficients account for all the control of flux down to, and including, the ring breakage enzyme (V), and Σc_{E_i} is indeed 1.0, within experimental error. However, in pyridoxine deficiency, the pyridoxal phosphate-dependent enzyme kynureninase (IV) assumes a measure of control of overall flux. This shows that there can be distribution of control lower down a pathway than a completely irreversible enzyme, which may, in

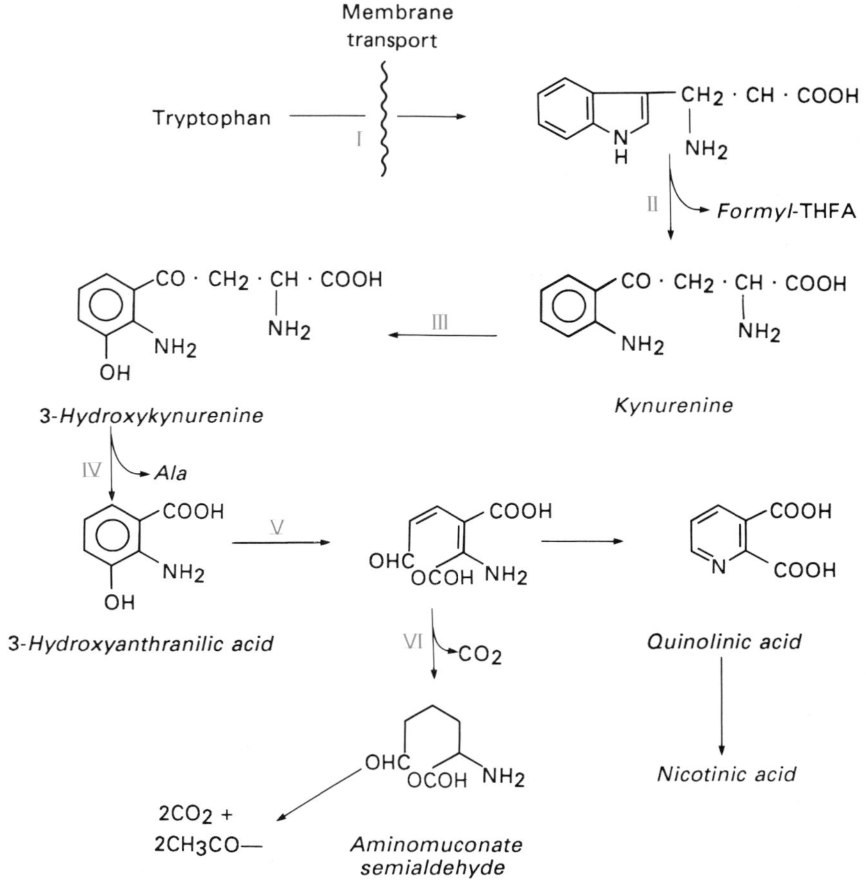

Figure 2.1. Tryptophan catabolism in animals. The numbers in orange refer to the enzyme steps listed in *Table 2.2*. Ala, alanine; THFA, tetrahydrofolic acid.

this example, be mediated by feedback inhibition by intermediary metabolites (13).

Finally, the bracketed figure opposite picolinate carboxylase in *Table 2.2* needs some comment. It was derived in terms of the flux through that branch of tryptophan metabolism alone. If it were added to the sum of earlier control coefficients, it would clearly break the summation rule of Equation 2.4, but the likelihood is that changes in the activity of this enzyme have no effect at all on the flux through TDO, and its global value is therefore probably approximately zero. It *will* have a redistributive effect on the flux through the other branches of the tryptophan degradation pathways, but this may well be entirely passive, and it does not automatically imply a large negative value of c_{E_i} in this branch. This complication is considered further under Section 4.2.

Table 2.2.   Flux control coefficients for tryptophan catabolism

	Treatment		
	None	Dexamethasone induction	Pyridoxine deficiency
Metabolic step[a]			
transport (I)	0.24	0.74	n.m.
	Σ 0.99	Σ 0.99	
TDO (II) + formamidase	0.75	0.25	n.m.
kynurenine hydroxylase (III)	<0.04	<0.04	n.m.
kynureninase (IV)	≃0	n.m.	0.28
3-OH-anthranilate oxidase (V)	≃0	n.m.	n.m.
picolinate carboxylase (VI)	[≃0.5]	n.m.	n.m.

Data derived from refs 12 and 13.
n.m. = not measured.
[a]The numbers in brackets refer to the steps in *Figure 2.1*.
Note that picolinate carboxylase is a misnomer, picolinate is only a dead-end
by-product of the reaction catalysed by this enzyme.

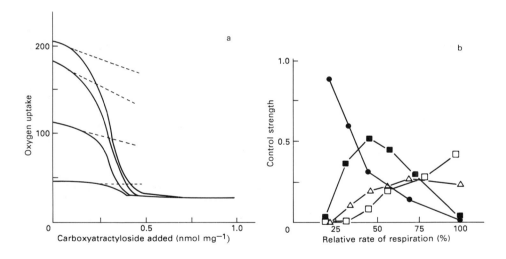

Figure 2.2.   Metabolic flux analysis of succinate oxidation by rat-liver mitochondria.
(a) Method of estimating the control strength of a single step (the adenine nucleotide
translocator) by drawing a tangent to each curve as the inhibitor concentration approaches
zero. The different respiration rates were obtained by varying the amount of added
hexokinase (+ glucose) to vary the rate of supply of ADP. **(b)** Variation in control strengths
at different respiration rates. □, dicarboxylate carrier; △, adenine nucleoside translocator;
●, proton leak; ■, hexokinase. For full details, see ref. 14.

3.3 Respiratory control of succinate oxidation by mitochondria

This is the most complete example of experimentally determined control strength
analysis yet published, and it involved a great deal of elegant work to determine

ΔJ for the limit as ΔE approaches zero for a number of mitochondrial transport and catalytic reactions (14). There is space here to show only one of these determinations, together with the profound change in the distribution of control on change from mitochondrial respiratory state 4 to state 3 (*Figure 2.2*). Two major points may be made: first that the sum of experimentally measured c_{E_i} is not 1.0 for any respiration state, which means that some control lies elsewhere. The discrepancy is not important, given that Equation 2.4a holds, because Σc_{E_i} is approximately 0.86, so that the unresolved components are minor. Moreover, it is the *change* in distribution of control as the mitochondria move from resting state to 100% oxygen uptake which is of particular interest.

The second point is that much of the control lies with processes that transfer reactants across the boundaries of the system, such as the succinate transporter or the adenine nucleotide translocator. This is discussed further at the end of this chapter.

4. Other applications of sensitivity analysis

4.1 Relative change sensitivity

This is a superficially similar fractional coefficient which is in use for systems not necessarily in a steady state; it is widely used in endocrinology (15). The coefficient s is defined as follows.

$$s_X^Y = (dY/dX) \cdot (X/Y) = d\ln Y/d\ln X \qquad [2.9]$$

Neither Y nor X need be fluxes; they could both be metabolic concentrations. In a steady state this latter use would have no meaning. There is also an intrinsic sensitivity coefficient s_{iX}^J where i is not a numbering subscript.

For example, $s_Z^{J_g}$ is the sensitivity of the glycolytic flux J_g to an external regulator Z (note the different use of the symbol Z in this notation compared with the sensitivity coefficient Z in Section 1). For the development and use of this form of analysis the reader is referred to the original literature (16,17).

4.2 Supersensitivity

Koshland *et al.* (18) have made a comprehensive study of fluxes in the citric acid cycle of *Escherichia coli*, in which the glyoxylate shunt operates, and isocitrate dehydrogenase (IDH) is regulated by reversible phosphorylation (*Figure 2.3*). The results are summarized in *Table 2.3*.

When the cells are grown on glucose, virtually no acetate carbon goes through the shunt, while on acetate about 30% of the citrate flux is shunted to malate. Koshland has pointed out (19) that this apparently makes the glyoxylate shunt enzymes *supersensitive* to nutrient changes, but in fact the response is a passive one to a 170-fold change in isocitrate concentration. This is due to simultaneous down-regulation of citrate synthase and IDH in response to glucose. In mechanistic terms, the response operates because IDH has a very low K_m for

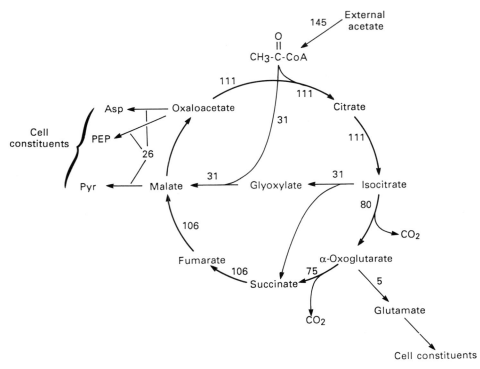

Figure 2.3. Partition of acetyl residues between various branches of intermediary metabolism during growth of *E.coli* on acetate. See text and *Table 2.3*. Asp, aspartate; Pyr, pyruvate.

Table 2.3. Citric acid cycle fluxes in *E.coli*

	Nutrient	
	Glucose	Acetate
Flux through citrate synthase (mmol $min^{-1} l^{-1}$)	22	111
Flux through IDH	22	80
Flux through glyoxylate shunt	<0.3	31
Relative activity of IDH	80%	20%
[Free isocitrate] μM	$\simeq 0.5$	95

K_m for IDH, 8 μM; for isocitrate lyase, 600 μM.
Data adapted from Koshland *et al.* (18).

isocitrate, and is consequently always saturated with its substrate, while isocitrate lyase has a much lower affinity, and the flux through it is always concentration-dependent. A not dissimilar phenomenon was found for acetoacetate metabolism in heart muscle (20), and the example of the picolinate and quinolinate branches of tryptophan catabolism (Section 3.2) is also relevant.

The latter example, in particular, demonstrates that extreme care is necessary in analysing branched pathways. Negative values of c_{E_i}, when an increase of $[E_i]$ causes a *decrease* in flux, certainly do occur (e.g. ref. 21), and are balanced by large positive values elsewhere; a very interesting example is discussed in (8). It is vital that care is taken to see that all observations relate to the same flux.

5. Discussion

There has been considerable controversy about metabolic control analysis, much of it misconceived. For a wide-ranging discussion the reader is referred to recent correspondence (22). The arguments can be summarized as follows.

(i) The analysis applies only to steady-state conditions. This is true, but purists forget how difficult it is even to analyse the steady state experimentally. Isotopic tracer kinetics have been studied almost exclusively under steady-state assumptions for almost 40 years, because any other approach is mathematically intractable, but have nevertheless provided many results of inestimable value.

(ii) The assumptions about an infinite source and an infinite sink, both of constant concentration, are unrealistic. This is a valid point, but in most *in vitro* experiments with single cells or cell organelles, care is taken to see that the primary substrate concentration is indeed constant. A chemostat is a more complex example of the same concept, and analysis of its behaviour is not always elementary. A more pertinent point is that the Kacser and Burns notation can lead to neglect of the transfer across the system boundaries, which may be non-enzymatic, but can nevertheless have a large element of control (see *Table 2.2* and *Figure 2.2*).

(iii) The mathematical derivation is true only for enzymes obeying Michaelis – Menten kinetics; it cannot be correctly applied to enzymes with non-hyperbolic kinetics, or to multi-enzyme clusters. The brief answer is that a complex response surface can always be 'linearized' by taking the increments over a very short interval, a proviso which was written into the original treatment. A more devious answer is that other applied mathematicians, notably statisticians, apply much more ruthless operations to their primary data, *because they believe that the major conclusions that they are then able to draw are still valuable.*

(iv) Equation 2.4 holds exactly only for 'point' concentrations of enzymes, that is, concentrations which can be neglected by comparison with those of the substrates; this assumption is made in the standard derivation of the Michaelis – Menten equation. This is very often not true *in vivo*; the inaccuracy does not matter with 'open' substrates—those that can be replenished from an infinite sink (11)—but only Equation 2.4a holds for 'conserved' substrates, that is, cofactors whose total amount in any cell compartment is fixed. The sequestration of reduced nicotinamide-adenine dinucleotide (NADH) by dehydrogenases in liver, for example, is quite easily detectable by fluorescence quenching measurements (23). The discrepancy in Equation 2.4 for NADH is not, in fact, large (11), but with other cofactors, particularly 'free' CoA, and possibly also with S-adenosyl methionine and tetrahydrofolate species, it may be very significant. Note that this effect is not catered for by 'moiety-conserved' analysis (24), which simply takes account of the fact that the sum of ATP together with ADP and AMP is a constant.

6. Summary and conclusions

There is little doubt that the treatments both of Kacser and Burns, and of Crabtree and Newsholme (16), are subsets of a much more elaborate analysis by Savageau (see ref. 22 and Section 7), but the latter is only intelligible to persons with considerable mathematical training. In spite of the possible limitations summarized above, metabolic control analysis can be recommended because it is simple to understand, (relatively!) simple to apply, and because it has predictive value (i.e. it can be used to design experiments from which new insights may be drawn). Familiarity with metabolic control analysis makes it clear that the important regulatory enzymes in a pathway often change with the environmental conditions, and that terms such as 'pacemaker', 'committed step', or 'flux-generating reaction' should only be used with circumspection—if at all.

7. Further reading

Savageau,M.A. (1976) *Biochemical Systems Analysis*. Addison-Wesley, Reading, MA.

8. References

1.   McMinn,C.L. and Ottaway,J.H. (1976) *J. Theor. Biol.*, **56**, 57.
2.   Ottaway,J.H. and McMinn,C.L. (1980) *FEBS Symp.*, **60**, 69.
3.   Kacser,H. and Burns,J.A. (1973) *Symp. Soc. Exp. Biol.*, **27**, 65.
4.   Heinrich,R. and Rapoport,T.A. (1974) *Eur. J. Biochem.*, **42**, 89.
5.   Westerhoff,H.V., Groen,A.K. and Wanders,R.J.A. (1984) *Biosci. Rep.*, **4**, 1.
6.   Rolleston,F.S. (1972) *Curr. Top. Cell. Regul.*, **5**, 47.
7.   Kacser,H. and Porteous,J.W. (1987) *Trends Biochem. Sci.*, **12**, 5.
8.   Westerhoff,H.V. and Arents,J.C. (1983) *Biosci. Rep.*, **4**, 23.
9.   Fell,D.A. and Sauro,H.M. (1985) *Eur. J. Biochem.*, **148**, 555.
10. Sauro,H.M., Small,J.R. and Fell,D,A. (1987) *Eur. J. Biochem.*, **165**, 215.
11. Ottaway,J.H. (1979) *Biochem. Soc. Trans.*, **7**, 1161.
12. Salter,M., Knowles,R.G. and Pogson,C.I. (1986) *Biochem. J.*, **234**, 635.
13. Stanley,J.C., Salter,M., Fisher,M.J. and Pogson,C.I. (1985) *Arch. Biochem. Biophys.*, **240**, 792.
14. Groen,A.K., Wanders,R.J.A., Westerhoff,R.V., Van der Meer,R. and Tager,J.M. (1982) *J. Biol. Chem.*, **257**, 2754.
15. Crabtree,B. and Newsholme,E.A. (1987) *Trends Biochem. Sci.*, **12**, 4.
16. Crabtree,B. and Newsholme,E.A. (1985) *Curr. Top. Cell. Regul.*, **25**, 21.

17. Crabtree,B. and Newsholme,E.A. (1978) *Eur. J. Biochem.*, **89**, 19.
18. Koshland,D.E., Walsh,K. and LaPorte,D.C. (1985) *Curr. Top. Cell. Regul.*, **27**, 13.
19. Koshland,D.E. (1987) *Trends Biochem. Sci.*, **12**, 225.
20. Ottaway,J.H. and McMinn,C.L. (1979) *Biochem. Soc. Trans.*, **7**, 411.
21. Saunderson,C.L. and Sauro,H.M. (1987) Abstracts of the American Society of Microbiology Conference on *Experimental and Theoretical Analysis of Metabolic Processes.* p. 16.
22. Discussion forum in *Trends Biochem. Sci.*, **12**, 216.
23. Sies,H., Tjiu,Ta.S., Brauser,B. and Bücher,T. (1972) *Adv. Enzyme Regul.*, **10**, 309.
24. Hofmeyr,J.-H.S., Kacser,H. and Van der Merwe,K.J. (1986) *Eur. J. Biochem.*, **155**, 631.

3

Regulation of ligand binding

'A theory has only the alternative of being right or wrong. A model has a third possibility: it may be right, but irrelevant.'

Max Eigen

1. Introduction

The evocation of a conformational change when a ligand binds reversibly to a protein underlies the whole of metabolic and physiological regulation. The whole concept of binding hormones and other extracellular active substances to receptors on the outer cell membrane (e.g. chemotactic substances for bacteria) implies a conformational change in the receptor that is in some way transmitted through the membrane (by mechanisms outside the scope of this book), to initiate an event either in the membrane itself, as with neutrophil bursts (1) or phospholipase A_2 activation (2), or the release of some second transmitter inside the cell. Even if this transmitter is a substance such as cyclic AMP (cAMP), which initiates a cascade ending in reversible covalent modification, the cell response still depends on a conformational change in the cAMP-sensitive receptor or protein kinase for the cascade to be put into effect.

Many of the external activators or inhibitors are quite large molecules, such as protein hormones or opioid peptides. Traditionally, however, regulation by ligand binding has come to mean the binding of small molecules (mol. wt < 1000) to proteins, and it is this restricted class of regulatory events with which this chapter primarily deals. The main theoretical analyses, and the important classes of ligands, end-product feedback regulators (3), for instance, were established almost 20 years ago, and there is not a very large volume of work, except in certain specialized fields, which is particularly recent. Thus I make no apology for treating the well-established theory in a slightly cursory way. Good up-to-date treatments are provided by Hammes and Fersht (see Section 6), and by Ricard (4).

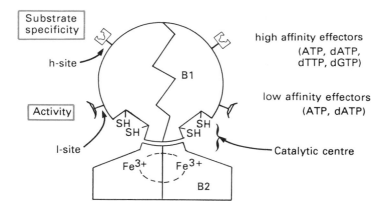

Figure 3.1.   Schematic diagram of the tetrameric structure of nucleoside diphosphate reductase (redrawn from ref. 5). The catalytic sites are shared between the B1 and B2 subunits, but the allosteric sites, which determine not only V_{max} (l sites), but also substrate specificity (h sites), are confined to the B1 subunit dimer.

2. The importance of quaternary structure

Ever since Monod *et al.* published the concerted transition model of cooperative ligand binding, based on the behaviour of haemoglobin, in 1963 (5), it has been generally assumed by biochemists that enzymes regulated by allosteric ligands must have a subunit structure. Many enzymes do indeed have at least two subunits, and it is difficult, although not impossible, to propose models for non-hyperbolic steady-state kinetics that do not involve subunit interaction, but the proposition itself always seemed inherently improbable. In fact, at least one very important regulated enzyme, the nucleoside triphosphate reductase of *Leichmanii leichmanii* and related bacteria, has been purified to homogeneity, and shown to consist of a single peptide chain with a molecular weight of 67 000. It does not aggregate, and it shows no sign of cooperativity, but its two allosteric binding sites not only regulate its affinity for the various ribonucleoside triphosphate (rNTP) substrates, but also for the prosthetic group cobalamin. In the absence of deoxynucleoside triphosphate (dNTP) effectors, the enzyme binds cobalamin so weakly that it is almost inactive (6). On the other hand, the more widespread NDP reductase, although it has two pairs of more or less identical subunits, does not behave at all like an enzymatic equivalent of haemoglobin. Its two catalytic sites are shared between the α and β subunits and the α subunits have two distinct allosteric sites (6,7) (see *Figure 3.1*). The positive high-affinity (h site) effectors simultaneously increase V_{max} and decrease K_m for the favoured NDP so effectively that reaction rate at low substrate concentrations can be increased 50- to 100-fold. The negative effector deoxyATP (dATP) favours conversion of the enzyme to an inactive dimer. In spite of this high degree of regulation by ligand binding, NDP reductase shows no homotropic cooperativity, and the curve of *v* against [S] is completely hyperbolic (6).

There are many other examples of complex responses to binding of effector ligands. Even when allosteric behaviour of enzymes is 'orthodox', it may still be difficult to elaborate any general principles, because the degree of variation between organisms can be very large. For example, yeast glyceraldehyde phosphate dehydrogenase shows concerted homotropic behaviour towards its substrate NAD^+, while the enzyme from rabbit muscle shows extreme negative cooperativity (8). A study of phosphofructokinase from *E.coli* provided strong justification for the Monod theory (9), but in Chapter 5 the extent of the divergences in the structure of this enzyme between prokaryotes and eukaryotes, and between the plant and animal kingdom, is stressed. It is still important to understand the concerted and induced-fit transition models, because they provide the only theoretical basis for the description of the behaviour of many polymeric enzymes, but the discussion here is brief, and an attempt is made to indicate lines along which future progress may occur.

2.1 The Adair and Hill equations

These are both phenomenological equations, that is they are mathematical descriptions of certain phenomena (e.g. oxygen molecules binding to haemoglobin, which they do quite accurately) in terms of constants with arbitrary relationships. Historically the Hill equation precedes the Adair equation, but can be derived from it. The Adair equation was derived (10) for the tetramer haemoglobin with four equivalent, but not independent, binding sites, each with an *intrinsic* (site) binding constant K_1, K_2, etc. It can be generalized to give the mean fractional saturation with the ligand $\overline{Y}$ (the bar used here and elsewhere in the text denotes the mean value for all the protomers).

$$\overline{Y} = \frac{\overline{\nu}}{n} = \frac{\sum\limits_{i=1}^{n} i\,\psi_i[L]^i}{n\left(1 + \sum\limits_{i=1}^{n} \psi_i[L]^i\right)} \qquad [3.1]$$

where $\overline{\nu}$ is the mean fractional saturation of the enzyme by the ligand L and $\psi_1 = K_1$, $\psi_2 = K_1 K_2$, ... $\psi_n = K_1 K_2 \ldots K_n$. If all the constants $K_1 \ldots K_n$ are equal (the sites are identical and independent of one another) Equation 3.1 reduces to

$$\overline{Y} = \frac{\overline{\nu}}{n} = \frac{K'[L]}{1 + K'[L]} \qquad [3.2]$$

the familiar hyperbolic isotherm.

Adair assumed that the ligand binds stepwise, whereas Hill had earlier (11) proposed an empirical equation based on the assumption that n molecules of ligand are bound in a single step. Then Equation 3.1 becomes

$$\overline{Y} = \frac{\overline{\nu}}{n} = \frac{\psi_n[L]^n}{1 + \psi_n[L]^n} \qquad [3.3]$$

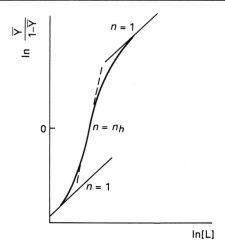

Figure 3.2.   Idealized Hill plot for an enzyme showing cooperativity. The Hill number is estimated as shown.

which can be transformed to

$$\ln[\overline{Y}/(1 - \overline{Y})] = \ln\psi_n + n\ln[L] \tag{3.4}$$

If the left-hand side of Equation 3.4 is plotted against $\ln[L]$, a straight line of slope n should be obtained. In fact, because ligand binding is not simultaneous, Hill plots are always curved (see *Figure 3.2*). The *Hill coefficient*, h, is the value of the slope of this plot at half-saturation with ligand. By deriving it from the Adair equation one can show that h cannot exceed n, the number of cooperating subunits, so experimental estimates of h can, in suitable cases, provide a lower limit for n. Hill himself, however, was careful not to claim any physical meaning for either of his constants, and h usually has a fractional value. Its chief use is to establish that cooperativity exists (i.e. non-equivalence or non-independence of binding sites). A *cooperativity index*, the ratio of $[L]$ values either for $Y_{0.8}/Y_{0.2}$ or $Y_{0.9}/Y_{0.1}$ can give the same information. For hyperbolic binding, the index values can be shown to be 16 and 81, respectively (see ref. 12 for further details).

2.2 Allosteric behaviour

Equations 3.1 and 3.3 were developed to describe multiple binding of a single ligand. Before summarizing the familiar concerted and sequential models of ligand binding, it may be helpful to give graphic examples of the ideal behaviour of enzymes showing allostery with a single ligand (homotropic) and more than one ligand, involving a second binding site (heterotropic behaviour), see *Figures 3.3* and *3.4*. Note that negative cooperativity is different from substrate inhibition (*Figure 3.5*), which can arise for several more complex reasons.

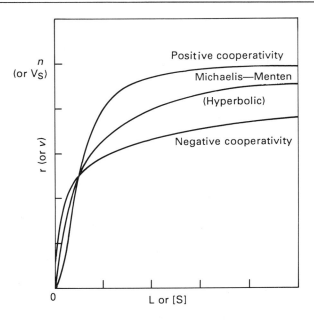

Figure 3.3. Michaelis–Menten, positive cooperativity and negative cooperativity relationships. The curves apply equally, given the assumptions discussed in Section 3, to pure ligand L or to substrate concentration [S].

2.3 The concerted and sequential (induced fit) models of ligand binding

Both these models depend ultimately on the hypothesis that ligand binding induces a change in the shape of the enzyme either locally or over the whole molecule, and also on the implicit assumption that regulated enzymes are at least dimers. We have seen that this idea is not universally true, but the models of Monod, Wyman and Changeux (13) and Koshland, Nemethy and Filmer (14) have had, and continue to have, enormous influence. There is no space here to discuss their derivation in detail; their similarities and differences are listed in *Table 3.1*. (For a complete treatment, see ref. 12 or Hammes in Section 6.)

For the concerted model, the fractional saturation for an n subunit protein is

$$\overline{Y} = \frac{K_A^*[S](1 + K_A^*[S])^{n-1} + LcK_A^*[S](1 + cK_A^*[S])^{n-1}}{(1 + K_A^*[S])^n + L(1 + cK_A^*[S])^n} \qquad [3.5]$$

where L is the ratio of conformations A and B (originally R and T) in the pre-equilibrium. The symbol L for ligand has been replaced by S. K_A^* is the microscopic (intrinsic) binding constant for the A conformation, K_B^* is the equivalent constant for the B conformation and c is the ratio K_B^*/K_A^*.

If L is zero, that is if the protein pre-exists in one state, the cooperativity vanishes. If c has a value of one, the binding is hyperbolic. However, if K_B^* is

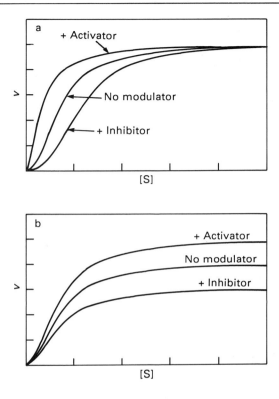

Figure 3.4. Expected effects of an allosteric ligand on a cooperative enzyme obeying the concerted transition model. (**a**) Effect on substrate affinity (K system); (**b**) effect on catalytic efficiency (V system).

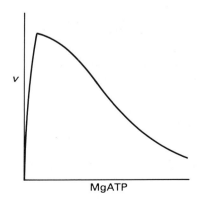

Figure 3.5. Inhibition of phosphofructokinase by its substrate MgATP at constant fructose-6-phosphate concentration. In this instance the inhibition at high substrate concentration is due to MgATP binding at a second, allosteric site.

Table 3.1.   List of differences between the concerted transition and simple sequence transition models of allostery

Concerted transition	Simple sequential transition
1. The protomers occupy equivalent positions.	No assumption necessary.
2. The free enzyme occurs in two different conformations.	The free enzyme exists in one conformation only.
3. The ligand can bind to both conformations.	As a simplifying assumption, only one of the two possible protomer configurations can bind the ligand.
4. Ligand binding shifts the pre-equilibrium existing between the two conformations of the enzyme.	Ligand binding to a protomer induces a conformational change in it, which is not propagated to the other protomers.
5. During the shift of the pre-equilibrium, symmetry is conserved (concerted transition).	In the simple model (loose subunit binding), symmetry need not be conserved, and the geometry of multiple subunit interactions is important.

zero, that is if the ligand binds to only one conformation, cooperativity is still present. Finally, by comparing Equation 3.5 with the Adair equation, Equation 3.3, it can be shown that in this model $K_4 > K_3 > K_2 > K_1$, so that cooperativity is necessarily positive (or $h > 1$).

It avoids confusion to know that the concerted model can be described in the symbolism of the simple sequential transition model, and is then known as 'tight subunit binding', but this does not mean that the model is different from that originally proposed by Monod *et al.* (5).

The corresponding equation for the sequential induced-fit model is somewhat more complex, and will not be quoted here. It is derived in terms of K_s, the association constant of ligand binding to conformation B, K_t, the notional equilibrium constant for the transition between the isolated A and B protomers, and three constants of interaction of the subunits, K_{AA}, which is that between the A subunits (usually set to 1, i.e. ΔG is zero), K_{AB} and K_{BB}. These constants are generally useful because they can be related, as seen here, to the free energies of association of the protomers. Even without postulates about non-equivalence of the ligand binding constants, the simple sequential transition model can generate negative as well as positive cooperativity, because there are no restraints on the sizes of K_{AB} and K_{BB}. In fact, for a dimeric enzyme

$$ h = 2/(1 + K_{AB}/\sqrt{K_{BB}}) \qquad [3.6] $$

Thus h can be >1 or <1.

2.3.1 Allosteric ligands

Both models of ligand binding can cope with heterotropic effectors: in the first case an inhibitor will stabilize the A conformation of the enzyme; in the second, it will diminish the interactions between AB and BB (cf. Equation 3.6).

3. Structural kinetics applied to allosteric enzymes

So far, the word 'substrate' has been avoided, because from Hill onwards all the treatments have referred to equilibrium binding, and it has to be assumed that the steady-state reaction velocity parallels the fractional ligand binding. In many cases this may be nearly true, but even so, this over-simplification obscures much interesting information. One must, in addition, always bear in mind that very few enzymes with sigmoidal velocity (or binding) curves have been shown to exhibit cooperativity rigorously (15).

Already in 1930, Haldane (16) had pointed out that the 'lock and key' theory of enzyme specificity is unsatisfactory, because a reversible enzyme cannot 'fit' both substrate and product at the same time. It is much more likely that the enzyme is complementary to the transition state, or if, for example there is more than one substrate, that it undergoes a series of conformational changes. The often-quoted example of hexokinase, which will not hydrolyse ATP, although water is a better phosphoryl acceptor than glucose, and is present in much higher concentration, is very compelling evidence for such changes. In fact, current theories of enzyme catalysis stress that the energy barrier at the transition state is lowered by releasing strain in the enzyme, but not the substrate, by a series of conformational changes *away from* the active site. In effect, the three-dimensional structure of an enzyme during a catalytic cycle must always be seen as a dynamic entity. It is attractive to think that changes in subunit interaction, even in polymeric enzymes without cooperativity, for example lactate dehydrogenase, contribute to an increase of catalytic activity in comparison with a monomeric enzyme.

By making a number of simplifying assumptions (4), in particular the ideal one that enzyme strain is completely relieved in the transition state, it is possible to predict, for example that enzymes showing negative cooperativity of binding may also show substrate inhibition (cf. *Figure 3.4*) in steady-state kinetics. More precisely, for a dimeric enzyme, it is possible to derive two equations, one for substrate binding

$$\overline{Y} = \frac{(\alpha_{AA}/\alpha_{AB})c + (\alpha_{AA}/\alpha_{BB})c^2}{1 + 2(\alpha_{AA}/\alpha_{AB})c + (\alpha_{AA}/\alpha_{BB})c^2} \qquad [3.7]$$

and one for the rate equation

$$\overline{V} = \frac{c + (\alpha_{AA}/\alpha_{AB})c^2}{1 + 2(\alpha_{AA}/\alpha_{AB})c + (\alpha_{AA}/\alpha_{BB})c^2} \qquad [3.8]$$

where c is equal to $K^*[S]$ and α_{AA}, α_{AB} and α_{BB} are substrate dissociation constants. By subtracting the equation for the ideal monomer

$$Y^* = V^* = c/(1 + c) \qquad [3.9]$$

from Equations 3.7 and 3.8, one can obtain an estimate of the effect of the dimerization on binding and on reaction rate. Equations 3.7 and 3.8 have identical denominators, but different numerators, and the two 'difference' equations are similar, but opposite in sign. This implies that a subunit interaction which enhances substrate affinity for an enzyme decreases the steady-state reaction rate, and vice versa. This is a very interesting prediction, because many allosteric ligands do indeed affect catalytic power rather than the K_m of enzymes (cf. *Figure 3.4a* and b), but they have always been neglected in the equilibrium ligand binding models of Monod and Koshland (even though the former referred to 'K' and 'V' enzymes). Until now there has been little that one could say about the mechanism by which such effectors operate.

Of course the simple treatment outlined above does not explain the complex behaviour of enzymes like ribonucleoside diphosphate (rNDP) reductase, but at least we get a hint that the conformational change, however it is transmitted from the allosteric sites, must affect the conformation of the transition state even more markedly than that of the substrate binding state, and this may be true also for many allosteric enzymes.

4. Enzymes with functionally different subunits

So far it has been assumed that each protomer has a catalytic site. However, there is an important class of enzymes regulated by reversible ligand binding for which this is not true.

4.1 Lactose synthase
This enzyme, better called galactosyl transferase, catalyses the reaction

$$UDP-Gal + Acceptor \longrightarrow Acceptor-Gal + UDP$$

where the acceptor can be the artificial substrate *N*-acetylglucosamine, or hexosamine residues on glycosylated proteins (e.g. fetuin). It is a monomeric enzyme (mol. wt $\simeq$ 47 000) which is found particularly in lactating mammary tissue and in milk. It binds tightly, but reversibly, to a smaller protein (mol. wt 17 000) which is one of the major components of milk (α-lactalbumin), and it will then use glucose as an acceptor, that is it is a lactose synthase (17). The data in *Figure 3.6* suggest that the modifier ('B' protein) does not bind sufficiently closely to the active site on the 'A' protein (enzyme) to displace polypeptide substrates, but that it causes a conformational change which favours glucose binding and diminishes that of free hexosamine. *Figure 3.6* also shows that the

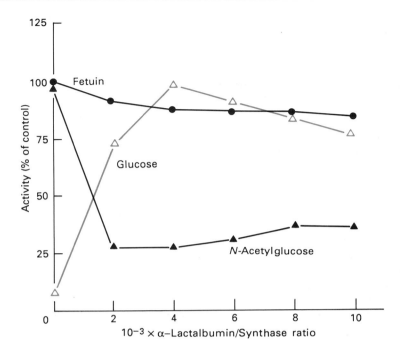

Figure 3.6.   Effect of adding α-lactalbumin ('B' protein) on the substrate specificity of galactosyl transferase. The ratio of 'B' protein to 'A' protein (enzyme) is very large, but the concentration of the latter in the assay solution is only $\simeq 10^{-8}$ M.

'A' protein can be titrated with the 'B' protein, until the ratio of the two is 1:1. To some extent, lactose synthase (see also Chapter 5) is a model which can help us to understand the way in which calmodulin works. The latter is a protein of similar molecular weight to the 'B' protein, and in some instances binds reversibly to the enzyme which it modulates.

   One can also ask if molecules as large as calmodulin, the Ca^{2+}-binding protein of quinate:NAD^+ oxidoreductase (see Section 2.3 of Chapter 6), or the 'B' protein can be called ligands, even though there is good evidence (18) that the last-named dissociates and re-associates during each catalytic cycle. Even more complex examples of reversible non-covalent association of large polypeptides are not infrequent. The dissociation of the α subunit of transducin after it has bound GTP, for example, and its association with the γ subunit of rod cell phosphodiesterase, which is described in Section 1 of Chapter 5, cannot be formally differentiated from the binding of small molecular weight ligands, particularly as the process goes to completion in a matter of milliseconds.

4.2 The protein kinases

These are among the best-known of the enzymes whose activity is masked by a protomer which dissociates when a regulatory ligand binds to it. They are not

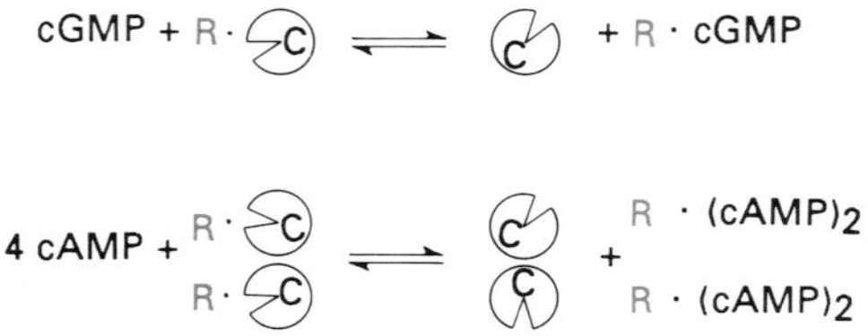

$$\text{cGMP} + \text{R} \cdot \overset{\frown}{\text{C}} \;\rightleftharpoons\; \overset{\frown}{\text{C}} + \text{R} \cdot \text{cGMP}$$

$$4 \text{ cAMP} + \begin{matrix} \text{R} \cdot \overset{\frown}{\text{C}} \\ \text{R} \cdot \overset{\frown}{\text{C}} \end{matrix} \;\rightleftharpoons\; \begin{matrix} \overset{\frown}{\text{C}} \\ \overset{\frown}{\text{C}} \end{matrix} + \begin{matrix} \text{R} \cdot (\text{cAMP})_2 \\ \text{R} \cdot (\text{cAMP})_2 \end{matrix}$$

Figure 3.7. Schematic diagram of the activation of cGMP and cAMP protein kinases. The active centres on the catalytic subunits (C) are partially masked by the regulatory subunits (R). Note that the regulatory subunit of cAMP-dependent protein kinase remains a dimer even after dissociation of the two catalytic monomers.

a homogeneous group of enzymes; for example cyclic GMP (cGMP)-sensitive kinase is a dimer $(\text{R} \cdot \text{C})$ while the cAMP-sensitive enzyme is a tetramer $(\text{R}_2 \cdot \text{C}_2)$ (see *Figure 3.7*). The conformational changes that occur when cAMP binds (19) can be summarized as follows. cAMP binds and forms a ternary complex $\text{R}_2 \cdot \text{C}_2(\text{cAMP})_n$ before the two types of subunit dissociate. The C subunits and cAMP each stimulate the dissociation of the other from their binding sites on the R subunits; complete saturation of the four regulatory sites with cAMP lowers the association constant between the R and C subunits by a factor of 10^4 (20). There is very high cooperativity between the two R subunits with h values of 1.6 – 1.8, which means that all four binding sites for cAMP have to interact with one another. There is also high cooperativity between the two catalytic subunits with respect to dissociation (19).

Since it is the protein binding site of the catalytic centre, not the substrate ATP site, that is masked by the R subunits, it is to be presumed that autophosphorylation of the type II R subunit (Section 1.1 of Chapter 4) takes place *before* the holoenzyme is activated, and it is therefore the rate of dephosphorylation by protein phosphatase 2B, *inter alia*, which determines the period during which re-association between R and C is hindered (for the importance of this 'delay' type of regulation, see Section 1.1 of Chapter 5).

4.3 Carbamoylphosphate – aspartate transferase (CAT)

This bacterial enzyme is highly regulated, both by feedback from an end-product, CTP, and by allosteric activation by ATP, and also by homotropic cooperativity with respect to the substrate aspartate. The latter has been successfully interpreted by the Monod model. The enzyme has been intensively studied for over 20 years (21), but new facts continue to come to light (22).

It differs from the protein kinases in that the regulatory subunits do not dissociate completely. Instead, there is an easily discernible relaxation of the molecule (*Figure 3.8*), which presumably allows for easier access of the substrates. If the C subunits *are* completely dissociated by treatment *in vitro*, they are up to 30 times more active than in the intact hexamer. This bears directly on what was said in Section 3 about the importance of disseminated strain in

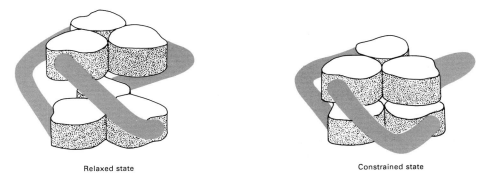

Relaxed state   Constrained state

Figure 3.8. Schematic structure of aspartate transcarbamoylase. There are two sets of catalytic trimers, with the three active sites in each trimer shared between the monomers (see text). The regulatory subunits (orange) are now known to be arranged in a left-handed helical configuration as shown.

improving catalytic efficiency. One can easily visualize that even in the relaxed state, the catalytic protomers are less able to release strain than they would be in free solution. One may surmise that in this enzyme, regulatory control has exerted more selection pressure than an increase in catalytic efficiency.

The three catalytic sites are shared between the three protomers (22). Such an arrangement is unusual, but it is found also in glutamine synthase (Section 2.2 of Chapter 4) and NDP reductase (Section 2), and it must allow quite small changes in conformation in a single protomer to have a noticeable effect on conformation at the active site, that is it facilitates 'fine tuning' of the enzyme activity.

5. Aggregating enzymes

It is possible to obtain mathematical models which display cooperativity as a result of reversible aggregation of protomers (see Hammes, Section 6), but enzymes which are known to aggregate do not seem to obey such simple rules. Several enzymes, for example NDP reductase and mammalian phosphofructo-kinase, are inactivated by oligomerization regulated by ligand binding. Acetyl-CoA carboxylase is so far in a class of its own, because the protomers (mol. wt $\simeq 400\,000$) are almost inactive. Polymerization is indefinite, to form a linear, fibrous molecule (*Figure 3.9*), with the protomers overlapping each other slightly at their ends (23). It is not known what conformational change is propagated along the protomers to make the catalytic sites active (presuming that they are not at the end of each peptide chain). Polyanions, of which citrate is the most familiar, favour polymerization, but it is not certain that citrate is the most important activator *in vivo*. The polymer is also inactivated by reversible phosphorylation (24).

Figure 3.9.   Arrangement of the protomers in a linear (fibrous) polymer of acetyl-CoA carboxylase. (Taken from ref. 23.)

6. Further reading

6.1 Models of cooperativity

Cornish-Bowden,A. (1976) In *Principles of Enzyme Kinetics*, Chapter 7. Butterworths, London.

Fersht,A. (1985) In *Enzyme Structure and Mechanism*, Chapter 10. Freeman, New York, 2nd edn.

Hammes,G.G. (1982) In *Enzyme Catalysis and Regulation*, Chapters 8 and 9. Academic Press, New York.

Koshland,D.E. (1970) In *The Enzymes*, Chapter 7. Boyer,P.D. (ed.), Academic Press, New York, 3rd edn.

6.2 Structural kinetics

Fersht,A. (1985) In *Enzyme Structure and Mechanisms*, Chapter 12. Freeman, New York, 2nd edn.

Ricard,J. (1985) In *Catalytic Facilitation in Organized Multienzyme Systems*. Welch,G.R. (ed.), Academic Press, New York, 3rd edn, p. 177.

7. References

1.   Babior,B.B. (1987) *Trends Biochem. Res.,* **12**, 241.
2.   Burgoyne,R.D., Cheek,T.R. and O'Sullivan,A.J. (1987) *Trends Biochem. Res.,* **12**, 332.
3.   Stadtman,E.R. (1970) In *The Enzymes*, Chapter 8, Volume 1. Boyer,P.D. (ed.), Academic Press, New York, 3rd edn.
4.   Ricard,J. (1985) In *Catalytic Facilitation in Organized Multienzyme Systems*. Welch,G.R. (ed.), Academic Press, New York, p. 177.
5.   Monod,J., Changeux,J.-P. and Jacob,F. (1963) *J. Mol. Biol.,* **6**, 306.
6.   Holmgren,A. (1981) *Curr. Top. Cell. Regul.,* **19**, 47.
7.   Thelander,L. and Reichard,P. (1979) *Annu. Rev. Biochem.,* **48**, 133.
8.   Kirchner,K. (1971) *Curr. Top. Cell. Regul.,* **4**, 167.
9.   Blangy,D., Buc,H. and Monod,J. (1968) *J. Mol. Biol.,* **31**, 13.
10.   Adair,G.S. (1925) *J. Biol. Chem.,* **63**, 529.
11.   Hill,A.V. (1910) *J. Physiol.,* **40**, iv.
12.   Cornish-Bowden,A. (1976) *Principles of Enzyme Kinetics*. Butterworths, London, p. 122.
13.   Monod,J., Wyman,J. and Changeux,J.-P. (1965) *J. Mol. Biol.,* **12**, 88.
14.   Koshland,D.E., Nemethy,G. and Filmer,D. (1966) *Biochemistry,* **5**, 365.
15.   Keech,D.B. and Wallace,J.C. (1985) In *Metabolic Regulation* (especially Addendum). Ochs,R.S., Hanson,R.W. and Hall,J. (eds), Elsevier, Amsterdam, p. 9.
16.   Haldane,J.B.S. (1930) *Enzymes*. Longmans, London.
17.   Fraser,I.H. and Mookerjea,S. (1976) *Biochem. J.,* **156**, 347.
18.   Takase,K. and Ebner,K.E. (1984) *Curr. Top. Cell. Regul.,* **24**, 51.

19. Hoppe,J. (1985) In *Metabolic Regulation*. Ochs,R.S., Hanson,R.W. and Hall,J. (eds), Elsevier, Amsterdam, p. 196.
20. Edelman,A.M., Blumenthal,D.K. and Krebs,E.G. (1987) *Annu. Rev. Biochem.*, **56**, 567.
21. Hammes,G.G. and Wu,C.-W. (1974) *Annu. Rev. Biophys. Bioeng.*, **3**, 1.
22. Schachman,H.K. (1987) *Biochem. Soc. Trans.*, **15**, 772.
23. Lane,M.D., Moss,J. and Polakis,S.E. (1974) *Curr. Top. Cell. Regul.*, **8**, 139.
24. Kim,K.-H. (1983) *Curr. Top. Cell. Regul.*, **22**, 143.

4

Regulation by reversible covalent modification

'He had a talent for appearing when he was not wanted, and a talent for disappearing when he was wanted.'

G.K.Chesterton

1. Phosphorylation

The earliest, and to date the most wide-spread system to be discovered, is reversible phosphorylation and dephosphorylation of one or more residues of the peptide chain of an enzyme. The acceptor group is usually an hydroxyl group, thus the residue phosphorylated is usually serine or threonine. In special instances tyrosine is also modified, but not the ring nitrogen of a histidine residue, which sometimes forms part of the active centre of phosphoryl-transferring enzymes. The amino acid residues are normally at specific sites, of the many available within the protein chain, and the introduction of a small group with a double negative charge clearly causes a significant modification of the conformation of the active centre (or sometimes of a regulatory protein) to take place. There is no way of predicting whether this change in conformation will lead to activation or inactivation, as *Table 4.1* shows (1,2).

The mechanism of activation or inactivation is also unpredictable; for example phosphorylation of phosphorylase kinase causes a large increase in V_{max}, while dephosphorylation of glycogen synthase reduces the K_m for the substrate UDP–glucose, as well as altering the affinities for the allosteric effectors glucose-6-phosphate (G-6-P), ATP, and inorganic phosphate (P_i). In addition, many regulated enzymes, of which glycogen synthase and cAMP-dependent protein kinase (cAMP–PK) itself are good examples, can be phosphorylated on several serine or threonine residues. Finally, the effectiveness of the phosphorylation may depend on other regulating factors. For example, the substrate P_i increases the affinity of phosphorylase b (the non-phosphorylated form) for the allosteric effector AMP, so that a combination of high concentrations

36

Table 4.1.   Some animal enzymes regulated by reversible phosphorylation

Enzyme	Principal tissues	Nature of kinase
Activation		
Phosphorylase	muscle, liver	specific (see next)
Phosphorylase kinase	muscle, liver	cAMP- and CaM-dependent
Tyrosine 5-monooxygenase	brain	cAMP- and CaM-dependent
Triglyceride lipase	adipose tissue	cAMP-dependent
Inhibition		
Pyruvate kinase (L-type)	liver	cAMP-dependent
Acetyl-CoA carboxylase	liver, adipose tissue	cAMP- and CaM-dependent
Glycogen synthase	muscle, liver	several (especially GSK3)
Pyruvate dehydrogenase	all tissues (and plants)	specific
Branched chain OADH	muscle, kidney	specific
HMG-CoA reductase	liver	specific (and PK-C and CaM-dependent)

HMG-CoA, hydroxymethylglutaryl coenzyme A; GSK3, glycogen synthase kinase 3; PK-C, protein kinase C.

of P_i and AMP would give a state in which phosphorylation (i.e. conversion to phosphorylase *a*) would not make much difference (3). It is unrewarding to pursue such details in a book of this kind. Attention has in any case focused on the controlling enzymes: originally on the kinases, but more recently also on the phosphatases, which are equally important.

1.1 The protein kinases of animal tissues

A recent, and excellent, review (4) discusses in depth 12 protein kinases that regulate enzymes, several more whose main function is to regulate protein synthesis, and it also mentions several others which have not yet been well characterized. Some of the kinases are specific for one protein substrate only, and indeed may form part of the structure of an enzyme complex, for example pyruvate dehydrogenase kinase and branched chain oxoacid dehydrogenase kinase (OADH). Others have a fairly broad specificity, and it is not always certain that the substrate on which they were first shown to have activity is the 'physiological' one. Protein kinases which are attached to membranes are not yet well characterized, but are growing in number. All these kinases are of course themselves regulated by ligand binding; the five ligands which are most widely distributed in animal tissues, namely cAMP, cGMP, calmodulin and Ca^{2+}, and diacylglycerol, are discussed in Chapter 5.

It is also remarkable that the activity of the soluble kinases themselves can be affected by phosphorylation, either by the ATP bound to the catalytic site (autophosphorylation) or sometimes by a separate kinase. It is noticeable, too, that it seems to be the exception rather than the rule for enzymes to be phosphorylated at only one site. The R_{II} subunit of cAMP − PK, for example, can contain phosphate on five different serine residues, only one of which is known to have regulatory significance (phosphorylation at Ser-95 reduces the rate of re-association with the C subunit, see Section 4.2 of Chapter 3), while

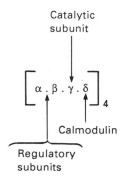

Figure 4.1.   Probable schematic structure of phosphorylase kinase.

the R_I variant can only be phosphorylated at two sites, neither of which is, with certainty, known to have significance (4). Differential location of the two subunits, for example R_I predominating in heart, with R_{II} predominant in adipose tissue, makes it likely that such differences do have physiological significance of some kind. The 'multisite' phosphorylation of glycogen synthase is much easier to explain, since each of the five possible phosphorylations reduces the activity of the enzyme to some degree (3). For details of these complex aspects of regulation the reader is referred to Section 3. Some details of the cooperativity of cAMP – PKs are described in Chapter 5.

1.1.1 Calmodulin-activated kinases

Although calmodulin (CaM) is known to regulate the biological activities of at least 25 biologically active proteins in animals (5), many of these are components of the cytoskeletal systems of cells or proteins with neuronal functions of brain cells (e.g. spectrin, tubulin kinase). Of the enzymes with clear-cut functions, two are concerned with the metabolism of cAMP and cGMP and are discussed in Chapter 5. Of the others, one is a kinase of fairly broad specificity which is known as *glycogen synthase kinase*, but which also activates two unrelated kinases in the brain; it is certainly not the only kinase which phosphorylates glycogen synthase. Another is *myosin light chain kinase* (MLCK) (4,6), which initiates contraction in smooth muscle cells, and in some way regulates the degree of tension in striated muscle. In non-muscle cells it may be concerned with motility and change of shape. Activation of MLCK is fairly straightforward; only CaM with four Ca^{2+} bound to it can bind reversibly; partially saturated CaM species cannot. The rate of inactivation when the ambient concentration of Ca^{2+} is lowered is about $1\,s^{-1}$, which is why direct involvement with myofibril contraction is only possible with slow-acting smooth muscle.

Finally, perhaps the best-known CaM-activated kinase is *phosphorylase kinase*. This enzyme is very complex, but shows similarities to protein phosphatase 2B (Section 1.2) and to quinate-NAD oxidoreductase (see Section 2.3 of Chapter 6). Its structure and the way it responds (4,7) are described here in some detail, because they illustrate the interactions between cAMP and CaM regulation. For other CaM-modulated enzymes, see Chapter 5.

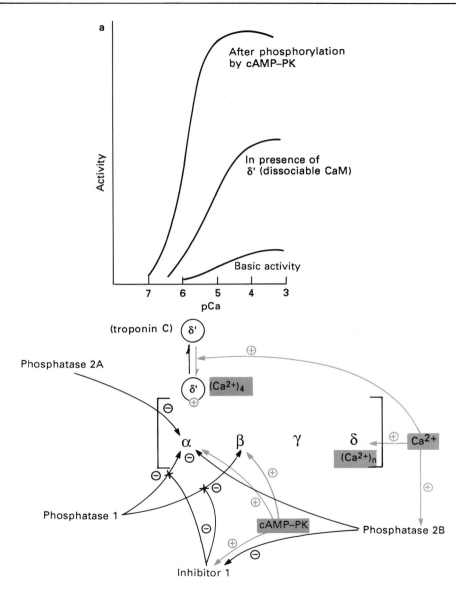

Figure 4.2.   Regulation of phosphorylase kinase activity. (**a**) Response of phosphorylase to pCa. The effects of dissociable CaM (troponin C) and of phosphorylation of the β subunit by cAMP – PK are clearly seen. (**b**) The effects of cAMP – PK are complementary on the kinase itself, but antagonistic when expressed through inhibitor 1.

The unit structure of phosphorylase kinase (repeated four-fold) is shown in *Figure 4.1.* Catalytic activity resides with the γ subunit. All the three other subunits, α, β and δ, modulate this activity, but α possibly less intensely than the other two.

Subunits α and β can be reversibly phosphorylated, the latter much faster than the former, both by cAMP – PKs and by autophosphorylation. The role that phosphorylated α plays in regulation is not clear. Phosphorylation of the β subunit increases both response to changes in $[Ca^{2+}]$ (Ca^{2+} concentration) (*Figure 4.2a*), and also decreases the K_m for the substrate, phosphorylase *b*.

Subunit δ is calmodulin; unlike MLCK, for instance, the peptide remains bound to the complex even when no Ca^{2+} is bound to it. Kinase activity is completely dependent on the presence of Ca^{2+}, as shown in *Figure 4.2a*. In addition to this permanently bound CaM, a second (extrinsic) CaM can associate with the α and β subunits (see *Figure 4.2b*), but only when this CaM is partially or completely loaded with Ca^{2+} as is more typical (Chapter 5). The α subunit exists in two isozymic forms, the α subunit found mainly in fast-twitch muscle, and a related α' subunit in slow-twitch and cardiac muscle. With the α subunit, the effect of the extrinsic CaM (δ') is additive to that of the intrinsic δ subunit with respect to Ca^{2+} bound, but δ' does not activate complexes that contain α', nor does it have any effect on the phosphorylated enzyme. This suggests that δ' interacts with the α and β subunits, which then affect the γ subunit, whereas it can be shown that the intrinsic CaM affects the γ subunit directly. An alternative modifier to δ' is troponin C, but its affinity is 100-fold less.

The complex interaction between the protein phosphatase and inhibitor 1, which is itself inhibited by cAMP-dependent phosphorylation, is also indicated in *Figure 4.2b*. This can affect the duration and intensity of the cAMP modulation. In summary, both cAMP – PK and Ca^{2+} activate phosphorylase kinase in different ways, and to different extents in different tissues, and the modes of interaction interact by means of changes in α and β subunit conformation. The changes in CaM conformation, on Ca^{2+} binding, are by now well known (8,9) (see also Chapter 5). It seems likely that the effect of cAMP is greater in most tissues, but that of Ca^{2+} is quicker—although it cannot be imagined that the activity of the kinase, much less that of phosphorylase itself, fluctuates in rhythm with changes in $[Ca^{2+}]$ in fast-twitch muscle.

1.2 Bacterial protein kinases

In bacteria, cAMP has several important functions, especially in Gram-negative bacteria where it regulates the initiation of transcription and translation of genes encoding carbohydrate permeases and catabolic enzyme systems. For this to occur cAMP has first to bind to a cAMP receptor protein (CRP) (10). No example is yet known of a bacterial protein kinase that is cAMP-activated (4). However, many protein kinases and phosphatases have been tentatively identified in bacteria, and two systems have been well studied. One of them, the regulation of bacterial isocitrate dehydrogenase, has been discussed in Chapter 3. For its associated kinase/phosphatase, one of the rare examples of a bifunctional protein, see Chapter 7; it appears to be modulated by allosteric metabolite binding (11). The other system, concerned with the regulation of carbohydrate uptake by inducer expulsion in Gram-positive bacteria, is illustrated in *Figure 1.1*. It also is regulated by metabolite feedback.

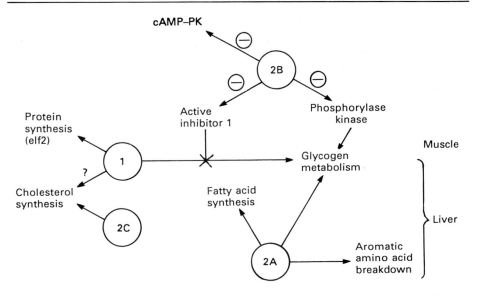

Figure 4.3.   Fields of effectiveness of the phosphoprotein phosphatases on intermediary metabolism.

1.3 Plant protein kinases

These are summarized in Chapter 6. One may be activated by cAMP (12), but it may be quite different from the animal cAMP – PK (4).

1.4 Phosphoprotein phosphatases

While it has always been clear that the existence of protein kinases presumes the existence of phosphoprotein phosphatase, just as adenyl cyclase is not a regulatory enzyme without the presence of phosphodiesterase, until recent years the possibility that phosphatases could be regulated has been neglected. Now, however, the picture is becoming clearer. There is a family of broad-specificity phosphoprotein phosphatases (1), with perhaps overlapping functions in different tissues. These have been named, on historical grounds, phosphatases 1, 2A and 2C (see *Figure 4.3*). It must be stressed that the enzymes do not have a tissue-specific distribution, phosphoprotein phosphatase 1, in particular, being found in brain, liver and adipose tissue as well as in muscle. Nevertheless, there appears to be some concentration of function as shown by the arrows in the diagram. On the other hand, phosphoprotein phosphatase 2B is almost inactive with the phosphorylated forms of the enzymes of metabolism summarized in *Figure 4.3*. Its substrates are regulatory proteins themselves, notably Type II cAMP – PKs (Section 1.1 and Section 2 of Chapter 5), phosphorylase kinase, and active inhibitor 1. It is also abundant in brain (calcineurin), although its function there is not so clear. Moreover, phosphatase 2B is itself regulated in a remarkable manner (*Figure 4.4*).

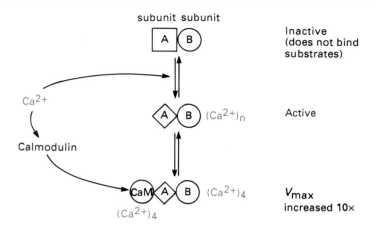

subunit subunit

Inactive
(does not bind
substrates)

Active

V_{max}
increased 10×

Figure 4.4.   Regulation of phosphoprotein phosphatase 2B.

The association both of the Ca^{2+}-binding subunit, which shows a 35% homology with CaM, and of loaded CaM itself, gives a very strong dependence on Ca^{2+}, and is formally not dissimilar from the arrangement in phosphorylase kinase (Section 1.1.1). The effect is of a Ca^{2+}-dependent phosphatase cascade, which can attenuate the cellular response to cAMP, while the converse is not true; that is cAMP has no regulatory effect on any of the phosphoprotein phosphatases, and, moreover, CaM activates the high K_m phosphodiesterase, probably the more important of the two enzymes which destroy cAMP and so limit its effectiveness (1).

2. Reversible nucleotidylation

2.1 Nomenclature

Covalent transfer of, for example, AMP to an acceptor by formation of a phosphodiester bond is adenylylation (because the complete free AMP group is *adenylic* acid). Similarly, transfer of UMP is uridylylation. Collectively, this process is nucleotidylation. Transfer of ADP – ribose (e.g. from nicotinamide) to an acceptor, with the linkage between ADP and the acceptor group through ribose, is called ADP – ribosylation. Although several examples of irreversible activation (e.g. of adenyl cyclase by cholera toxin) or inactivation (e.g. of elongation factor 2 by diphtheria toxin) are known, no example of *reversible* modulation of an enzyme by ribosylation has yet come to light. ADP – ribosylation in nuclei, as a method of regulation, is just beginning to attract interest (13).

2.2 The regulation of bacterial glutamine synthase

This is the only example of its type so far known, but it exhibits a number of features that are relevant to other systems, and an atomic model of the

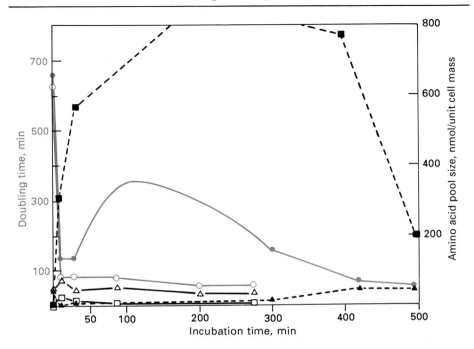

Figure 4.5.   Response of a normal strain (open symbols) and a mutant strain (filled symbols) of *Salmonella typhimurium* to an NH_4^+ load (at zero time). The mutant strain was unable to inhibit glutamine synthase by adenylylation (see *Figure 4.7*). Orange ($\bigcirc$, $\bullet$) curves, doubling time of culture; $\triangle$ and $\blacktriangle$, glutamate pool sizes; $\square$ and $\blacksquare$, glutamine pool sizes.

(unregulated) enzyme obtained by X-ray crystallography has recently been published (14). The mechanism of regulation is therefore described in some detail. The synthase catalyses the reaction of glutamate (Glu) with ammonium ions in an ATP-linked reaction to give glutamine.

$$Glu + NH_4^+ + ATP \longrightarrow Glu \cdot NH_2 + ADP + P_i$$

This reaction is coupled to that of an ϵ-amino transferase so that glutamine reacts with α-oxoglutarate (αOG) to give two molecules of glutamate

$$Glu \cdot NH_2 + \alpha OG \longrightarrow 2\ Glu$$

The importance of regulating these two steps is shown in *Figure 4.5*, which shows the effect of deregulating the first step by genetic manipulation (15). Although an enormous amount of glutamine is synthesized in response to an NH_4^+ load, there is very little increase in growth rate. The glutamate concentration falls almost to zero, and most of the glutamine that was synthesized has been lost

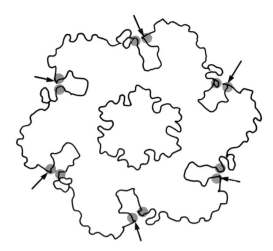

Figure 4.6. Plan view of the hexagonal arrangement of the subunits of a GS molecule from *S.typhimurium*. Six of the 12 subunits are shown; a similar hexagon lies directly underneath with the active centres (shown in orange) staggered. The locations of Tyr-397, the adenylylation sites, are indicated by the arrows. (Derived from ref. 12.)

to the extracellular medium, whence it is only slowly reabsorbed. *Figure 4.5* shows the effects of lack of regulation very clearly.

Glutamine synthase (GS) has 12 subunits, each with a molecular weight of 51 600 daltons, arranged in two hexagonal rings. Each catalytic site is shared between two adjacent subunits, which is unusual, but see Chapter 3, Section 4.4. A specific tyrosyl residue (Tyr-397) on each subunit, which also lies on the interface between subunits, can be adenylylated from ATP by a transferase usually abbreviated as ATase (*Figure 4.6*). Blocking the tyrosyl residue inhibits the corresponding active centre. The adenylylation is reversible, but in physiological conditions a separate reaction operates in which the adenyl group is transferred to P_i to give ADP and pyrophosphate (PP_i) (16).

Both catalytic sites, for adenylylation (AT_a) and de-adenylylation (AT_d), reside on the same single polypeptide chain (17). Unless the two activities could be switched, the result would simply be futile cycling, conversion of ATP and P_i to ADP and PP_i. The switch mechanism is provided by a link protein, simply called PII. This protein consists of four small subunits (mol. wt 11 000), each of which has a specific tyrosine residue which can accept a uridyl residue, with UTP as donor (*Figure 4.7*). PII has no catalytic activity. PII (or PII_o, the unmodified form) is an obligatory activator of ATase *a*. PIID (PII_m, the completely uridylylated form) is on the other hand an obligatory activator of ATase *d*, the site which catalyses the phosphorolysis of $(adenyl)_n$ – GS, and hence the release of synthase activity from inhibition. PIID is reconverted to PII by hydrolysis, as shown in *Figure 4.7b*; probably the UMP-transferase and the hydrolase activities also reside on a single protein (UR/UT) (16). It is not

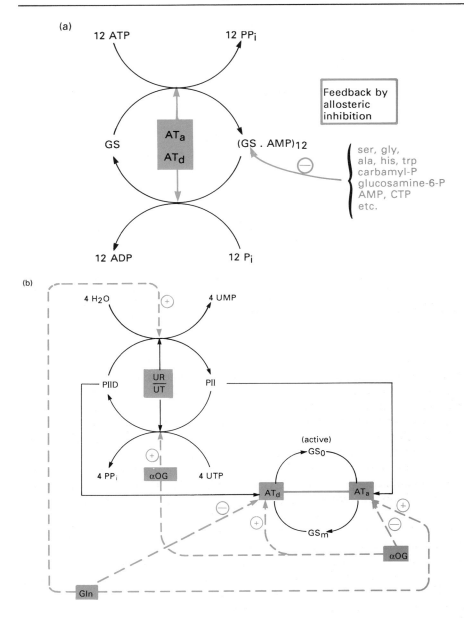

Figure 4.7.   Regulation of bacterial GS. (**a**) Transfer of an adenyl residue from ATP, catalysed by adenyl transferase (AT_a). The same bifunctional protein can also catalyse transfer of the adenyl residues from the synthase to P_i. The catalyst, then known as AT_d, is located at a different site on the transferase. Both AT_a and AT_d have an absolute requirement for a specific form of the regulatory protein PII, see (b). (**b**) Interconversion of PII and PIID, catalysed by the bifunctional protein uridyl transferase (UR/UT). The opposing roles of Gln and αOG in regulating both the ratio of PII to PIID, and the rates of activation and inactivation of GS itself, are indicated in orange.

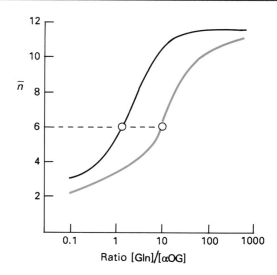

Figure 4.8.   Effect of the uridylylation of PII on the activity of GS. These curves were obtained from experiments on permeabilized bacterial cells. They show the [Gln]:[αOG] ratio necessary to maintain GS at half-activity (i.e. $n = 6$) when both cascade cycles (*Figure 4.7b*) are operating (black curve), and when only the adenylylation cycle is effective (orange curve).

clear whether hybrid tetramers with partial uridylylation are able to bind with perhaps reduced efficiency to both the ATase sites.

It is known, on the other hand, that the 12 sites of GS itself, although they may interact, are capable of independent catalysis. Thus the fraction of full catalytic potential that is available is related in a roughly linear fashion to n, the average number of adenylylated subunits per molecule of GS. In most physiological conditions (adenyl)$_{12}$–GS is completely inactive (16).

Both glutamine, the product of the GS reaction, and αOG, one of the substrates, are allosteric effectors at three sites, namely UT/UR, the AT_a site and the AT_d site. In each case they operate in opposite senses. ATP is also a positive allosteric effector of UT and ATase, while many nitrogen-containing small molecules, part of whose amino or ring nitrogen is derived from glutamine, are feedback inhibitors of adenylylated subunits of GS. In all, about 40 such effectors are known, operating on various parameters of the enzyme complex (16,18). The overall effect is to make bacterial GS respond very sensitively to the energy and nutritional status of the cell, while the amplification—that is the change in the ratio of [Gln]:[αOG] that is required to obtain 50% activation of GS (i.e. $n = 6$) is theoretically very considerable. *Figure 4.8* shows that experimentally, in permeabilized cells (19), operation of the PII (uridylylation) cycle lowers the ratio of [Gln:αOG] needed to obtain 50% activation by a factor of 10.

In addition to these controls, the amount of GS protein in *Escherichia coli* cells is subject to genetic regulation, and PII is an important modulating factor in this complex system (*Figure 4.9*).

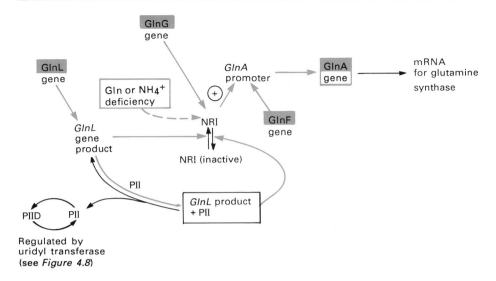

Figure 4.9. Genetic regulation of the synthesis of GS in bacteria. Four genes are involved in the regulation of the overall amount of the enzyme protein. Transcription of the *GlnA* gene provides the mRNA that codes for the enzyme, but the promoter region of the gene requires two factors before transcription occurs. One is a second gene, *GlnF*, or a product of it. The second is a protein, called NRI, which recognizes a deficiency of glutamine or ammonium ions. This is a fairly straightforward derepressor mechanism. NRI is coded for by a third gene, *GlnG*. The intervention of a complex between the product of a fourth gene, *GlnL*, and PII converts NRI into an active form. The *GlnL* gene product itself, on the other hand, catalyses the conversion of NRI (inactive) to NRI (active). The formation of the (*GlnL* gene product)–PII complex is reversible, so that the fraction of NRI in the active form is effectively determined by the ratio between PII, which switches glutamine synthase off, and PII(UMP)$_4$, which switches it on. This in turn depends on the intracellular concentrations of glutamine and αOG (see *Figure 4.7b*).

3. Further reading

Current Topics in Cellular Regulation, Volume 27 (1985). A Symposium in honour of E.R. and T.C.Stadtman, containing many invaluable reviews of covalent modifications of enzymes in animal and bacterial cells.
The Enzymes, Volumes 17 (A and B) (1986) and 18 (A and B) (1987) Boyer,P.D. and Krebs,E.G. (eds), 3rd edn. Both volumes are devoted to reversible phosphorylation of proteins, mainly enzymes. The A volumes contain more general articles, while the B volumes are given over to detailed reviews of specific kinases and phosphatases.

4. References

1. Cohen,P. (1985) *Curr. Top. Cell. Regul.,* **27**, 23.
2. Krebs,E.G. (1986) In *The Enzymes*, Volume 17A. Boyer,P.D. and Krebs,E.G. (eds), Academic Press, New York, 3rd edn, p. 3.
3. Martin,B.R. (1987) In *Metabolic Regulation—A Molecular Approach.* Blackwell Scientific, Oxford, p. 232.

4. Edelman,A.M., Blumenthal,D.K. and Krebs,E.G. (1987) *Annu. Rev. Biochem.,* **56**, 567.
5. Wang,J.H., Pallen,C.J., Sharma,R.K., Adachi,A.-M. and Adachi,K. (1985) *Curr. Top. Cell. Regul.,* **27**, 419.
6. Stull,J.T., Nunnally,M.H. and Michinoff,C.H. (1987) In *The Enzymes*, Volume 17A. Boyer,P.D. and Krebs,E.G. (eds), Academic Press, New York, 3rd edn, p. 114.
7. Picket-Gies,C.A. and Walsh,D.A. (1986) In *The Enzymes*, Volume 17A. Boyer,P.D. and Krebs,E.G. (eds), Academic Press, New York, 3rd edn, p. 396.
8. Tanaka,T. and Hidaka,H. (1980) *J. Biol. Chem.,* **255**, 11078.
9. Anderson,W.B. and Gopalakrishna,R. (1985) *Curr. Top. Cell. Regul.,* **27**, 455.
10. Rickenberg,H.V. (1974) *Annu. Rev. Microbiol.,* **28**, 353.
11. Nimmo,H.G. (1985) In *Metabolic Regulation*. Ochs,R.S., Hanson,R.W. and Hall,J. (eds), Elsevier, Amsterdam, p. 215.
12. Ranjeva,R. and Boudet,A.M. (1987) *Annu. Rev. Plant Physiol.,* **38**, 73.
13. Gaal,J.C., Smith,K.R. and Pearson,C.K. (1987) *Trends Biochem. Res.,* **12**, 129.
14. Almassy,R.J., Janson,C.A., Hamlin,R., Xuong,N.-H. and Eisenberg,D. (1986) *Nature,* **323**, 304.
15. Kustu,S., Hirschman,J. and Meeks,J.C. (1985) *Curr. Top. Cell. Regul.,* **27**, 201.
16. Stadtman,E.R. and Ginsburg,A. (1974) In *The Enzymes*, Volume 10A. Boyer,P.D. (ed.), Academic Press, New York, 3rd edn, p. 755.
17. Rhee,S.G., Park,C.R., Chock,P.B. and Stadtman,E.R. (1978) *Proc. Natl. Acad. Sci. USA,* **75**, 3138.
18. Garcia,E. and Rhee,S.G. (1983) *J. Biol. Chem.,* **258**, 2246.
19. Mura,U., Camici,M. and Gini,S. (1985) *Curr. Top. Cell. Regul.,* **27**, 233.

5

Signals, transducers and modifiers

'When I use a word, said Humpty Dumpty . . . , it means exactly what I want it to mean, neither more nor less.'
Lewis Carroll

1. Introduction

The word *signal* is one of the most imprecise in the English language. It has the connotation of generalized information transfer on which no action need be taken, as with most broadcast radio signals. It can more specifically mean that action *may* become necessary, as with a flashing light at a road junction, but a radio signal which will set off a detonator to explode a bomb is both specific and mandatory. A host of synonyms has grown up, such as 'second messenger', 'transducer', 'modifier', 'transmitter'—even 'magic spots' (1)—and it may be worthwhile to examine one or two examples in the hope of defining their usage more closely. Neurochemists have already appropriated some terms, so we shall start with an example from this field.

Figure 5.1 shows the mechanisms by which light energy, falling on receptors in rod cells of the retina, is transformed into a chemical change and amplified until it shuts many Na^+ gates in a membrane and initiates a nervous impulse (2,3). It demonstrates all but one of the regulation mechanisms with which this book is concerned. Conversion of light energy into a nervous impulse is called, quite properly, *transduction*; in the diagram there are two transductions. The first occurs when a photon falls on the retinal-binding protein, rhodopsin, and causes a change in the configuration of a second protein (transducin). The second occurs with the transitory fall in concentration of cyclic GMP (cGMP), which initiates a propagatable voltage change in a nerve axon. Both these processes are membrane-bound and are not enzymatic, so they will not be discussed further here. The important point is that both are fairly described as transformations of one form of energy into another. It is consequently confusing to name

49

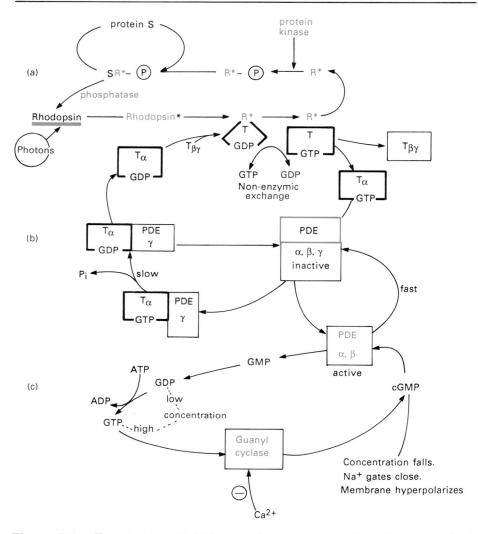

Figure 5.1.   Transduction of light impulses into membrane polarization by vertebral retinal rod outer segments. **(a)** Activation of membrane-bound rhodopsin by light. The activated rhodopsin (rhodopsin* or R*) then catalyses a conformation change in several hundred molecules of transducin (T), leading to the release of $T_\alpha \cdot$GTP (see *Figure 5.2*). **(b)** Inactive (membrane-bound?) PDE loses a masking subunit, PDE_γ, which associates with $T_\alpha \cdot$GTP. The masking subunit slowly inactivates $PDE_{\alpha\beta}$ as $T_\alpha \cdot$GTP converts to $T_\alpha \cdot$GDP. **(c)** PDE is able to bring about a transient reduction in the concentration of cGMP. The latter is necessary to hold open Na$^+$ gates in the membrane. cGMP is re-formed by guanyl cyclase, which is slowly regulated by Ca^{2+}.

mechanisms by which purely *chemical* energy is transmitted from one regulating system to another transduction (e.g. hormone receptor $\longrightarrow$ cAMP-linked protein kinase $\longrightarrow$ activated enzyme), even when the term 'chemical transduction' is used.

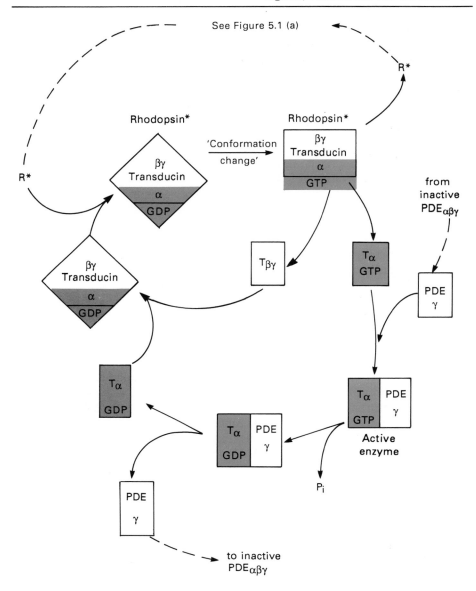

Figure 5.2. Detail of the transducin cycle (stage b of *Figure 5.1*). Activated rhodopsin causes a conformation change in $T_{\alpha\beta\gamma}$, which lowers its affinity for guanine nucleotides, and leads to replacement of GDP by GTP (whose concentration is higher). T_{α} escapes, and picks up PDE_{γ}. The association continues until the GTP slowly hydrolyses to GDP.

Activated rhodopsin is a catalyst, even if not an enzyme in the classical sense (see ref. 4); a single molecule of rhodopsin can modify about 500 molecules of transducin before it is inactivated. Secondly, the complex inactivation process is mediated by a protein kinase/phosphatase pair which is not activated by effectors; its function here is not amplification, but interval limitation. This is also a possible function of protein phosphorylation/dephosphorylation in more familiar enzymatic pathways.

The protein transducin (3) is neither a catalyst nor a concentration-dependent effector; it is basically a link, in the mechanical engineering sense, but this analogy cannot be pushed too far; a large reservoir of inactive transducin – GDP which can be converted into transducin – GTP is needed if the activation of phosphodiesterase (PDE), and the consequent initiation of the nerve impulse, can proceed at the necessary rate. The mechanism of activation and inactivation of transducin is summarized in *Figure 5.2*. Both the original hydrolysis of GTP and the indirect inactivation of PDE are interval-limiting processes, which ensure that the drop in cGMP concentration is transient. This is an important factor in many (but not all) enzyme regulation mechanisms, and one which is easily overlooked. Incidentally, a fast inactivator of PDE must exist, in addition to the cycle shown in *Figure 5.1*, because the whole transduction process is over, and the rod is ready to respond to another photon, in about 4 s. There is considerable interest in transducin at present, because it is a model for the 'G' proteins of the adenyl cyclase complex (see Section 2.1). In addition, the α subunit is structurally and functionally similar to the Tu elongation factor of bacterial protein synthesis, and to the p21 (mol. wt 21 000) product of mammalian *ras* oncogenes (3,5).

Finally, the cGMP part of the cascade (6) depends on the maintenance of a high concentration (60 μM total; 6 μM free) of this nucleotide during the resting state, by means of a guanyl cyclase whose activity is modulated by Ca^{2+}. The steady-state concentration of the latter depends on an inward leak and an outward pump. The perturbation of cGMP concentration is intense, because PDE is a very active enzyme, with a turnover number of approximately 4200 s^{-1}. The activation of PDE by removal of a masking subunit is not unusual; it is, after all, the mechanism of activation of cAMP-activated protein kinase, but it is not common for the removal to depend on the binding of the masking subunit to another protein.

1.1 Nomenclature

Before discussing nomenclature, let us look briefly at a completely different system: mammalian lactose synthase, which was described in Section 4.1 of Chapter 3. In spite of an apparent concentration dependence, which reflects the finite concentration of B protein needed to attain 1:1 equivalence, the effect is all or none: a single molecule of A protein is either converted to lactose synthase, or it is not. There is no concentration-dependent sequential conformation change, as in some kinds of ligand-binding effector modulation (Chapter 3). The B protein is therefore best described as a *modifier*.

Returning to the rod cell transduction system of *Figure 5.1*, we can make some progress in codifying the various components. Starting from the bottom of the diagram, there is no reason to suppose that [Ca^{2+}] changes very rapidly in this type of cell; thus Ca^{2+} is a modulator, rather than a trigger. The concentration of cGMP, on the other hand, does fall sharply and transiently and it is this which initiates the hyperpolarization of the membrane. Clearly, it is the *change* in [cGMP] that is the signal, rather than cGMP itself, and this is true for cAMP, Ca^{2+}, fructose-2,6-bisphosphate, and several other chemical substances whose sole function is to transfer information between one site in a cell and another. Put most simply, they are *transmitter* substances, but because this word is used extensively by neurophysiologists, biochemists seem reluctant to make use of it. Some of these transmitters, cAMP for instance, appear to be very old, because they are also found in fungi and prokaryotes, but the information that they transmit may be very different between one phylum and another (cf. Section 2.1), because they have been adapted to perform different functions. It is therefore misleading to call the compounds themselves signals, and, although the distinction may seem pedantic, it is of considerable help in understanding metabolic regulation. In this book the word 'signal' is reserved for an event which occurs on the outside of the cell membrane, and which therefore need not be described in detail.

Moving upward through *Figure 5.1*, we can see that although the release and inactivation of the α subunit of transducin is complex, its function is simple: it is a link molecule or *modifier*. It is not a modulator, because the response of the receptor, PDE, is not concentration-dependent, but is all-or-none, like that of the lactose synthase A protein. Compare this with calmodulin (Section 2.4.1), and in particular *Figure 5.4*.

Finally, the function of rhodopsin—other than that of initiating the chemical part of the transduction—is to be an *amplifier*, because it is 'switched on' for a much longer time than it takes the energy of the photon to be absorbed. This is a function which in the context of this book is much more frequently carried out by a protein kinase. However, the delay function operated by protein kinase/phosphatase in rod cells (top of *Figure 5.1*) is a perfectly valid part of a chemical amplification system, and is discussed as such by Chock *et al.* (7). In Chapter 1, Section 2.1, the practical amplification to be expected from one-stage and two-stage amplifiers is also discussed.

2. Major intracellular transmitters and modulators

2.2 Cyclic AMP and adenylate cyclase

Cyclic AMP is, in animals, a transmitter between signal receptor sites on the exterior surface of the cell membrane and regulatory enzymes of metabolism, via the amplifier cAMP-sensitive protein kinase. In bacteria, cAMP has quite a different function, that of modifying the transcription of specific genes in response to the lack of glucose in the external environment (8), while in the eukaryotic amoeba *Dictyostelium* cAMP is an *external* (chemotactic) signal between identical cells, initiating the aggregation process (9). In both these cell types the cyclase, although undoubtedly regulated, is not linked to a 'signal-sensitive' receptor site on the external membrane.

The animal system is particularly complex, because the receptor site is embedded on the outside of the membrane, while the cyclase is embedded on the inside, and the two are not functionally connected. Instead, linkage depends on the presence of two 'G' proteins which require GTP for activity (10,11). There are two, because typically each cyclase molecule can be stimulated by a ligand such as a β-adrenergic agonist or adrenocorticotropic hormone, or inhibited by an α-adrenergic agonist, muscarine or an opioid, binding to a separate receptor site.

The G proteins are functionally and structurally similar to transducins (5). For example they have an α and βγ subunit structure, and hypotheses about their mode of action have been based on what has been learnt about transduction in rod cells, as in *Figure 5.1*. Indeed the most popular hypothesis at present is an almost exact copy of the rhodopsin – transducin – PDE system, that is the α subunit is presumed to bind to and affect the cyclase itself, while the βγ subunits become detached and do not return until the GTP bound to the α subunit has been hydrolysed to GDP (10). However, the system cannot be as simple as this, as there are two G proteins, which can be distinguished; cholera toxin transfers ADP – ribose from NAD^+ to G_a, permanently switching on the cyclase, whereas *Pertussis* toxin ADP – ribosylates G_i, permanently inhibiting it. The chief objections to the simple hypothesis are

(i)   The kinetics of the response of the catalytic protein (C), both to loading of the receptor sites and to changes in [GTP] are simple and Michaelian in nature, unlike the complex response of protein kinase to cAMP, where a dissociation step *does* occur.

(ii) A $GTP \cdot \alpha_a \cdot C$ complex (active) can be isolated from membranes, but it has never been possible to find a $GTP \cdot \alpha_i \cdot C$ (inactive) complex.

(iii) If free βγ subunits inhibit the cyclase by favouring the reformation of $\alpha_a \cdot \beta\gamma$ (i.e. inactive G_a), then activation of other membrane-bound G proteins leading to release of βγ subunits, for example those related to phospholipase C, should switch off adenylate cyclase; this has not been shown to occur.

A review with alternative models of the activation/inactivation process has been published elsewhere (12).

As adenyl cyclase will remain switched on for an appreciable time after the stimulating receptor site is vacated, a maximal concentration of cAMP can be reached without saturation of the receptor sites (cf. the amplification provided by the rhodopsin system, *Figure 5.1*). However, the intracellular cAMP concentration depends also on its rate of hydrolysis by PDE. In fact, the approach to a new steady state can be quantified by the equation

$$d[E]/dt = K - k[E] \qquad [5.1]$$

where [E] is the concentration of effector, K is the rate of its formation, which usually modulates so rapidly that it can be regarded as constant, while the rate of destruction is proportional to the concentration of [E]. On integration Equation 5.1 gives

$$[E]_{\text{steady state}} = K/k\,(1 - e^{-kt}) \qquad\qquad [5.2]$$

Thus the new steady-state concentration, which may itself be relatively transient, is determined by the ratio K/k, but the speed at which it is obtained depends *only* on the rate of destruction (i.e. on k). This equation was first developed for predicting the kinetics of changes in concentration of inducible proteins in animals (13). In animal tissues, there are two PDE isoenzymes, one of low K_m ($\sim 1\ \mu$M) and one of high K_m ($\sim 100\ \mu$M) (14). The latter is present in much higher concentrations, and is the more important in hydrolysing cAMP (and cGMP) to the 5'-nucleotide. It is also the species which is activated by calmodulin, which in effect damps down the intracellular response to an external stimulus. In brain, calmodulin also stimulates adenyl cyclase (15), but this is not a general phenomenon.

2.2 Cyclic GMP

An important role for cGMP is established in rod vision as we have seen but there is no simply defined general role for this cyclic nucleotide (16). A cGMP-sensitive protein kinase is well-known (see Chapter 3, Section 4.2), but its general physiological function is also unclear (17).

2.3 ATP and AMP

These two compounds differ from the others discussed in this chapter, in that they are normal metabolites, and the normal intracellular concentration of ATP is much higher than that of all the other compounds discussed here ($\sim 1-5$ mM). Its concentration in animal cells does not normally vary very much, being regulated in cytosol (but not in mitochondria) by adenylate kinase (see Equation 5.3), and in muscle by creatine or arginine kinase, and also by the phosphorylation potential.

ATP is best known as an allosteric inhibitor of phosphofructokinase (see *Figure 5.3*), and as an activator of nucleoside diphosphate reductase (see Chapter 3, Section 2), and of several bacterial enzymes such as glutamine synthase (see Chapter 4, Section 2.2) and aspartate transcarbamoylase (Chapter 3, Section 4.3). It is necessary to be certain about the inhibiting species; for example $MgATP^{2-}$, but not ATP^{4-}, inhibits citrate synthase, which is located in mitochondria, where the only important species is ATP^{4-}. 'ATP inhibition' of citrate synthase in animal mitochondria is therefore a non-event. In yeast and prokaryotes, the concentration of ATP can be rather more variable, but its constancy in animal cells implies that effective modulation of enzymes such as phosphofructokinase depends on variations in concentration of antagonists of ATP inhibition.

The concentration of AMP is much lower than that of ATP, and less often

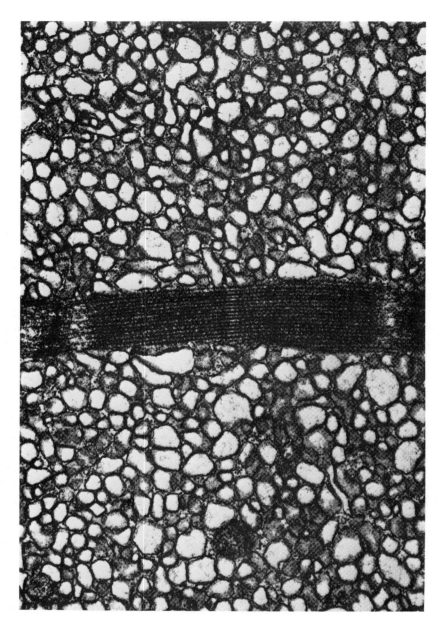

Figure 5.3.   The photograph (taken with permission from ref. 21) shows a single fibre from a very fast twitch muscle, with its associated sarcoplasmic reticulum, needed for the production of very rapid changes in [Ca^{2+}]. One mitochondrion can be seen at bottom centre.

measured directly. In cytosol it may be roughly estimated (again the question of species is important) by using the equilibrium constant (K_{eq}) ($\simeq 1$) of the adenylate kinase reaction

$$AMP + ATP \rightleftharpoons 2 \, ADP \qquad\qquad [5.3]$$

Good estimates of total ATP and ADP in heart muscle (18) are 5.2×10^{-3} and 1.0×10^{-3} mol kg^{-1} tissue, respectively. This gives an apparent value for [AMP] of 0.2×10^{-3} mol kg^{-1}. However, about 80% of the ADP in muscle is thought to be bound to actomyosin, and when correction is made for this, the value of [AMP] falls to 0.8×10^{-5} mol kg^{-1}, which is not much greater than that of cAMP at peak times. In liver the higher value for [AMP] is more likely to be true, but the uncertainties, particularly those arising from binding to enzymes, are considerable.

AMP is a powerful activator of phosphorylase b, and in some circumstances it is possible that, as a regulator, it is almost as important as conversion to phosphorylase a (Chapter 4, Section 1). It is also an inhibitor of fructose-1,6-bisphosphatase (19), and it used to be thought that it promoted glycolysis and inhibited glycogenolysis in liver. However, changes in carbohydrate flux in liver do not correlate well with changes in [ATP] and [AMP] (20), and it is likely that fructose-2,6-bisphosphate (see Section 2.5 below) is a more important modulator in this tissue.

2.4 Calcium and calmodulin

Calcium ions activate in two ways—by liganding directly to enzymes, and through calmodulin and similar proteins which are discussed below. The calpains, the Ca^{2+}-inhibited guanyl cyclase of rod cells (see Section 1), protein kinase C (Section 2.6), protein phosphatase 2B (see Chapter 4, Section 1.2), and, in plant cells, quinate:NAD$^+$ oxidoreductase (see Chapter 6, Section 2.3), are all examples of enzymes directly modulated by Ca^{2+}. The last two examples are complex, in that a dissociable Ca^{2+}-binding protein is also involved.

Animal cells have an active plasma membrane Ca^{2+} pump, a sequestering/pumping mechanism in endoplasmic or sarcoplasmic reticulum, and a long-term reservoir for Ca^{2+} in mitochondria. Nevertheless, one must have reservations about rapid signalling (triggering) by Ca^{2+} in all but specialized cells. The concentration of Ca^{2+} can be expressed in the same way as [H$^+$], that is as pCa, by analogy with pH. The concentrations of Ca^{2+} that modify enzyme activity range from 0.1 to 2 μM (pCa 7 – 5.8), which is in the same concentration range as H$^+$, and there are several intracellular Ca^{2+}-binding proteins other than calmodulin which must buffer changes in pCa, just as Brønsted bases buffer pH changes. Even in plant cells which are photosynthesizing, in which a rapid exchange of H$^+$ and Mg^{2+} occurs through chloroplast membranes, there is some scepticism about the importance of pH changes in regulating plant enzymes (see Chapter 6, Section 2.4). One may therefore doubt whether rapid changes in pCa occur in animal cells. This conclusion is strengthened by the fact that

Table 5.1.   Enzymes regulated by calmodulin[a]

Enzymes	Effect	Effect on 'target'
'High K_m' cNMP phosphodiesterase	+	
Protein phosphatase 2B	+	usually −
Brain adenylate cyclase	+	
Phosphorylase kinase	+	+
Myosin light chain kinase	+	
Multifunction protein kinase (kinase II)	+	some +, some −
Plant protein kinase	+	+

[a]Sources: refs 17,24 – 25.

in muscle cells where rapid triggering by Ca^{2+} *does* occur, the specialized structures necessary to bring this about are very extensive. *Figure 5.3* shows the overwhelming proportion of sarcoplasmic reticulum (21), in relation to the contractile fibril, necessary to bring about a 10 ms contraction/relaxation in a very fast twitch muscle.

Unfortunately, the latest review of Ca^{2+}-buffering (22) avoids any quantitative calculations of intracellular pCa buffering and the speed of pCa changes in non-muscle cells.

2.4.1 Calmodulin

Calmodulin (CaM), with a molecular weight of approximately 16 000, binds up to four Ca^{2+} ions in equivalent sites. As Ca^{2+} is bound, more and more hydro-phobic regions are exposed in the molecule (23). It operates in two ways: directly, by binding to enzymes, or indirectly through a CaM-dependent multi-functional protein kinase (PK II), which is identical with one of the glycogen synthase kinases, or the protein phosphatase 2B described in Chapter 4. The major enzymes regulated by CaM are listed in *Table 5.1*.

The relationship between varying pCa and the activity of CaM-modulated enzymes is complex, and depends on whether the CaM is permanently bound to the enzyme. Phosphorylase kinase and protein phosphatase 2B, in which this is so, have been discussed in Chapter 4; in other enzymes, CaM can dissociate. With PDE, at least, the dissociation constant of $CaM(Ca^{2+})_4$ is as much as 10^6 less than that of CaM (17,26). If the value of pCa is about seven, unless CaM is in excess, the species existing are the unmodulated enzyme E and $E - CaM(Ca^{2+})_4$, that is no intermediate $E - CaM(Ca^{2+})_{1-3}$ states exist; each enzyme molecule is either unmodulated or fully activated. This is assisted by a very rapid association of free E with $CaM(Ca^{2+})_4$. This implies a bi-state (on – off) control *for each individual enzyme molecule*. The very rapid and tight binding of $CaM(Ca^{2+})_4$ means, however, that in the cytosol there are also very few $CaM(Ca^{2+})_{1-3}$ species; the CaM molecules will also be either unloaded or fully loaded, and the free $[Ca^{2+}]$ in the solution will remain low, even when Ca^{2+} has been entering it. The extent to which the PDE reaches its full activity will depend on the *fraction* of the total enzyme molecules which have bound loaded

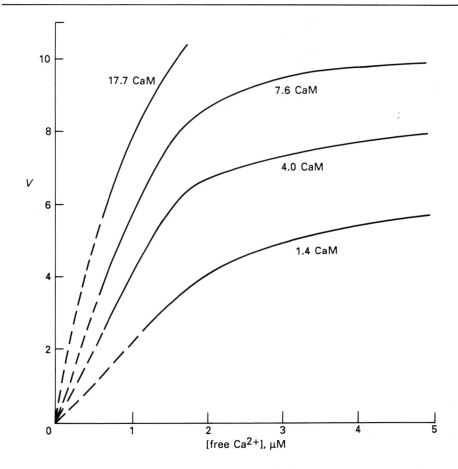

Figure 5.4. Velocity – activator curves for protein phosphatase 2B. The abscissa shows the *free* [Ca^{2+}]. Each curve represents the activity in the presence of a fixed concentration of CaM, shown above each curve in micromolar units. Redrawn from ref. 25.

CaM, which in the final resort depends on *how much* Ca^{2+} has entered the solution. *Figure 5.4* illustrates the relationship between [Ca^{2+}] and activity.

2.5 Fructose-2,6-bisphosphate

2.5.1 Range of activities

Fructose-2,6-bisphosphate (F-2,6-BP), discovered in 1981, is found in micromolar concentrations in many animal cells and in most plants and in yeasts, but not in prokaryotes. It appears to have no other function than that of a regulating ligand. It affects a very small number of enzymes rather intensively.

(i) Phosphofructokinase (PFK1). This enzyme is structurally rather variable. *Bacterial* PFK1 is a tetramer with a subunit molecular weight of 40 000, each

Table 5.2.   Relative $[L]_{50\%}$ for hexose bisphosphates on PFK1

	F-2,6-BP	F-1,6-BP	G-1,6-BP
Liver	1	1000	3000
Muscle	1	10	80

The data refer to the relative concentration of allosteric ligand, $[L]_{50\%}$, which produces 50% activation of the isoenzymes of PFK1 found in the two tissues. G-1,6-BP, glucose-1,6-bisphosphate.

subunit having a catalytic site; it is unaffected by F-2,6-BP. The *plant* enzyme, which is found only in chloroplasts, is also unaffected. The *yeast* enzyme is an octomer with four α (regulatory) and four β (catalytic) subunits. F-2,6-BP is a potent stimulatory ligand, lowering the K_m for fructose-6-phosphate (F-6-P) and increasing its V_{max}. The *animal* enzyme is different again; it is a tetramer with a molecular weight of approximately 340 000, composed of different arrangements of three types of subunit (M, L and F). This means that different tissues have isoenzymes with qualitatively different properties. It seems likely that the subunits of 80 000 molecular weight arose by gene duplication from the 40 000 molecular weight subunits (cf. bacteria, above), and that one of these two sites lost catalytic activity. Thus the allosteric site also binds F-2,6-BP and glucose-1,6-bisphosphate (G-1,6-BP), but with varying affinity in different tissues (27). This is illustrated in *Table 5.2*. This means that in liver, F-2,6-BP is always the major effector of PFK1 (other than AMP and ATP, see Section 2.3), but in muscle, since it is present in 1000-fold lower concentration than fructose-1,6-bisphosphate (F-1,6-BP) and is only 10 times more effective, it is unimportant as a stimulator of glycolysis. In addition, the liver isoenzyme has a great and unique tendency to aggregate, and the aggregated form has a lower K_m for its substrate F-6-P. F-2,6-BP favours aggregation of the enzyme and ATP opposes it (27).

(ii) Fructose-1,6-bisphosphatase (FBPase 1). F-2,6-BP is an inhibitor of this enzyme, found in plant cytosol, yeast, and animals (mainly in liver). The plant chloroplast enzyme is regulated in a different way (see Chapter 6, *Figure 6.2*). The enzymes from the sources just listed are not identical, but it seems likely that in each instance F-2,6-BP binds at an allosteric site which overlaps the catalytic site, so that inhibition is competitive, and therefore not always complete. An interesting point is that the effector converts an 'oversquare' curve of v against [S] to a sigmoidal one (28) (see *Figure 5.5*). In other words, it converts negative cooperativity to positive cooperativity. This is not orthodox allosteric behaviour (see Chapter 3, Section 2.3 and *Figure 3.4*) although it could be explained by the simple sequential transition model of ligand binding. AMP is synergistic with F-2,6-BP and inhibits non-competitively, suggesting that it binds at a separate site (29).

(iii) Pyruvate kinase. In *Trypanosoma brucei*, F-2,6-BP is equally as effective as an activator as F-1,6-BP for pyruvate kinase, but not in any other organism known at present (27).

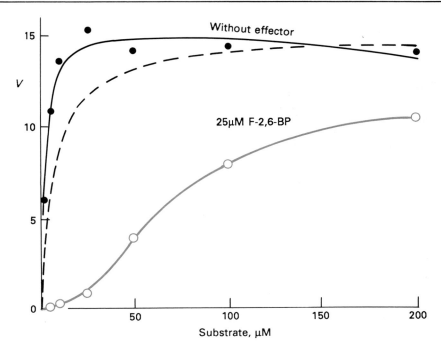

Figure 5.5.   Inhibition of FBPase 1 by F-2,6-BP. The upper curve shows the behaviour of the enzyme in the absence of modulator. The curve is 'oversquare', indicating negative cooperativity. The dashed line shows an arbitrary hyperbola for comparison. The lower orange curve is sigmoidal, indicating a positive cooperative response to substrate in the presence of F-2,6-BP. This concentration gives maximal inhibition.

(iv) Pyrophosphate F-6-P phosphotransferase. This enzyme catalyses the reversible reaction between F-6-P and pyrophosphate (PP_i)

$$F\text{-}6\text{-}P + PP_i \rightleftharpoons F\text{-}1,6\text{-}BP + P_i \qquad [5.4]$$

It is found in some bacteria and in high concentration in the cytosol of almost all plants. Its activity is almost entirely dependent on F-2,6-BP. As it is an equilibrium enzyme, it is difficult to say that it controls gluconeogenesis or glycolysis, but for its possible role in regulating the balance between starch and sucrose synthesis in leaves, see refs 30 and 31.

Both PFK1 and FBPase 1 are phosphorylated by a cAMP-dependent PK (cAMP – PK), but this is not thought to be physiologically important (27).

2.5.2 The regulation of fructose-2,6-bisphosphate concentration

The compound is synthesized from F-6-P and ATP by a specific kinase (PFK2) and hydrolysed back to F-6-P by a specific phosphatase (FBPase 2). In liver, both catalytic sites are found on a single bifunctional protein (see Chapter 7,

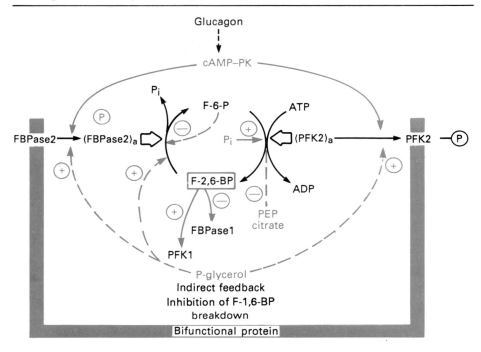

Figure 5.6. Regulation of F-2,6-BP concentration in liver. The fact that FBPase 2/ PFK2 is a bifunctional enzyme is indicated diagrammatically. A single phosphorylation by a cAMP – PK will thus both activate FBPase 2 and inactivate PFK2. In addition, phosphoglycerol is an important allosteric and active site modulator.

Section 1.1). Activity is regulated in liver, but not in any other animal tissue, by cAMP – PK; phosphorylation inactivates PFK2 and activates FBPase 2. Additionally, F-6-P, the substrate for PFK2, inhibits FBPase 2, while glycerol-3-phosphate and a number of other triose phosphates inhibit PFK2 or stimulate FBPase 2, or both (32). The effects on F-2,6-BP concentration are shown diagrammatically in *Figure 5.6*.

In yeast, a cAMP – PK phosphorylates PFK2 and *activates* it—the opposite effect from that in liver, although the two PFK enzymes are largely homologous. There is little FBPase 2 activity in yeast, and it is not certain that there is a bifunctional protein. Thus although cAMP may be a 'hunger' signal in bacteria, in yeast it acts to promote glycolysis.

Some authors (e.g. ref. 32) have described F-2,6-BP as a signal which integrates metabolic information. Reference particularly to the control of PFK2/FBPase 2 in liver shows, however, that it is these enzymes, together with FBPase 1, which respond to, and integrate, the metabolic state of the cell. The small molecule is a variable 'flag' or link, which transmits the integrated metabolic 'status' of the cell to responsive metabolic enzymes. A similar conclusion may be drawn from considering the completely opposite effects of cAMP – PKs in liver and yeast; to think of cAMP as a 'hunger' or 'low blood sugar' signal is

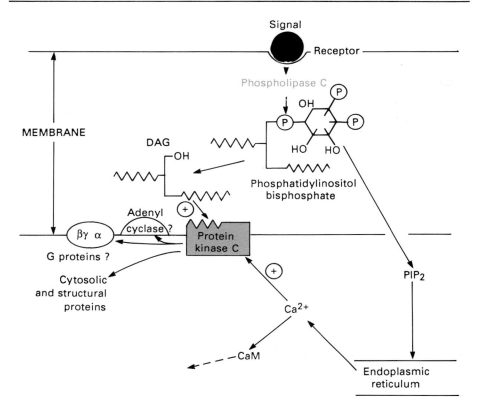

Figure 5.7. Stages in the activation of PK-C. The enzyme is depicted as bound to the inner surface of the plasma membrane.

to over-simplify the response and to misinterpret the mechanism. It is one catalytic protein which has evolved, in particular cells, a fast or slow integrating response. Comparison with the status of bacterial glutamine synthase (Chapter 4, Section 2.2), which is never either fully active or fully inactive, is useful here.

2.6 Triphosphoinositol, diacylglycerol and protein kinase C

Protein kinase C (PK-C) has a wide range of substrates, many of which are also substrates for cAMP – PKs. Nevertheless, its characteristic substrates are regulatory structural proteins which are not enzymes, so it is only briefly treated in this book. It is activated by Ca^{2+}, phospholipids and diacylglycerol (DAG); the inter-relation between the three activators is very complex (17). An attractive hypothesis is that extracellular signals which cause release of DAG result in translocation of PK-C from cytosol, where it is presumably inactive, to membranes. At the membrane surface it can be activated by Ca^{2+}, DAG and a native phospholipid of the membrane.

The extracellular signals referred to above appear to cause phospholipase C to hydrolyse phosphatidylinositolbisphosphate, PIP_2, a minor membrane constituent, to triphosphoinositol (IP_3) and DAG (33). There are no restrictive requirements for fatty acid residues in DAG as an activator of PK-C, so it cannot formally be distinguished from DAG as a metabolic intermediate, but presumably these particular DAG molecules remain in the vicinity of the membrane. PIP_2, on the other hand, is thought to travel to the endoplasmic reticulum and stimulate Ca^{2+} release (*Figure 5.7*); if this is true, it is a genuine transmitter or flag. Protein kinase C is found in greatest concentration in brain and platelets, in lowest concentration in adipose tissue and in yeast, and neither the release of Ca^{2+} nor protein phosphorylation have any relation to muscle contraction.

There is little doubt that both phospholipase C and its substrate are located within the plasma membrane, but Chabré (34) has recently re-emphasized that only integral proteins can be truly said to form part of the membrane, and that others, such as the G proteins, are within the aqueous environment of the cytosol, although they are often strongly adsorbed to the inner surface of the membrane. This view is still strongly disputed and a full discussion is beyond the scope of this book, but it is as well to remember that the precise location of so called 'membrane phenomena' is in fact often rather uncertain.

3. Further reading

Most of the material in this chapter is so new, and opinions are developing so rapidly, that there is no text that can be recommended. The following articles may be of interest:

Klein,W.L. (1984) *Curr. Top. Cell. Regul.*, **24**, 129.
Koshland,D.E. (1985) In *Metabolic Regulation*. Ochs,R.S., Hanson,R.W. and Hall,J. (eds), Elsevier, Amsterdam, p. 1 and p. 190.

4. References

1.   Nierlich,D.P. (1978) *Annu. Rev. Microbiol.*, **32**, 393.
2.   Stryer,L. (1987) *Sci. Am.*, **257**, 32.
3.   Fung,B.K.-K. and Navon,S.E. (1987) *Biochem. Soc. Trans.*, **15**, 39.
4.   Wessling-Resnick,M., Kelleher,D.J., Weiss,E.R. and Johnson,G.L. (1987) *Trends Biochem. Res.*, **12**, 473.
5.   Bourne,H.R., Masters,S.B. and Sullivan,K.A. (1987) *Biochem. Soc. Trans.*, **15**, 35.
6.   Stryer,L. (1986) *Annu. Rev. Neurosci.*, **9**, 87.
7.   Chock,P.B., Rhee,S.G. and Stadtman,E.R. (1980) *Annu. Rev. Biochem.*, **49**, 813.
8.   Gerisch,G. (1987) *Annu. Rev. Biochem.*, **56**, 853.
9.   Rickenberg,H.V. (1974) *Annu. Rev. Microbiol.*, **28**, 353.
10.   Gilman,A.G. (1984) *J. Clin. Invest.*, **73**, 1.
11.   Gilman,A.G. (1987) *Annu. Rev. Biochem.*, **56**, 615.
12.   Levitski,A. (1987) *FEBS Lett.*, **211**, 113.
13.   Shimke,R.T. (1969) *Curr. Top. Cell. Regul.*, **1**, 77.
14.   Klee,C.B., Crouch,T.H. and Richman,P.G. (1980) *Annu. Rev. Biochem.*, **49**, 489.
15.   Livingstone,M.S., Sziber,P.P. and Quinn,W.G. (1984) *Cell,* **37**, 205.

16. Goldberg,N.D. and Haddox,M.K. (1977) *Annu. Rev. Biochem.,* **46**, 823.
17. Edelman,A.M., Blumenthal,D.K. and Krebs,E.G. (1987) *Annu. Rev. Biochem.,* **56**, 567.
18. Wilson,D.F., Nishiki,K. and Erecinska,M. (1985) In *Metabolic Regulation.* Ochs,R.S., Hanson,R.W. and Hall,J. (eds), Elsevier, Amsterdam, p. 77.
19. Kustu,S., Hirschman,J. and Meeks,J.C. (1985) *Curr. Top. Cell. Regul.,* **27**, 201.
20. Katz,J. and Rognstad,R. (1985) In *Metabolic Regulation.* Ochs,R.S., Hanson,R.W. and Hall,J. (eds), Elsevier, Amsterdam, p. 95.
21. Rosenbluth,J. (1969) *J. Cell Biol.,* **42**, 534.
22. Carafoli,E. (1987) *Annu. Rev. Biochem.,* **56**, 395.
23. Anderson,W.B. and Gopalakriskna,R. (1985) *Curr. Top. Cell. Regul.,* **27**, 455.
24. Cohen,P. (1985) *Curr. Top. Cell. Regul.,* **27**, 23.
25. Kauss,H. (1987) *Annu. Rev. Plant Physiol.,* **38**, 47.
26. Huang,C.Y. and King,M.M. (1985) *Curr. Top. Cell. Regul.,* **27**, 437.
27. Van Schaftingen,E. (1987) *Adv. Enzymol.,* **59**, 315.
28. Van Schaftingen,E. and Hers,H.-G. (1981) *Proc. Natl. Acad. Sci. USA,* **78**, 2861.
29. Gottschalk,M.E., Chatterjee,T., Edelstein,I. and Marcus,F. (1982) *J. Biol. Chem.,* **257**, 8016.
30. Preiss,J. (1984) *Trends Biochem. Res.,* **9**, 24.
31. Cseke,C., Balogh,A., Wong,J.H., Buchanan,B.B., Stitt,M., Herzog,B. and Heldt, H.W. (1984) *Trends Biochem. Res.,* **9**, 533.
32. Hers,H.-G., Hue,L. and Van Schaftingen,E. (1985) In *Metabolic Regulation.* Ochs,R.S., Hanson,R.W. and Hall,J. (eds), Elsevier, Amsterdam, p. 222.
33. Taylor,C.W. and Merritt,J.E. (1986) *Trends Pharmacol. Res.,* **7**, 238.
34. Chabré,M. (1987) *Trends Biochem. Res.,* **12**, 213.

6

Regulation of enzymes in plants

Are plants human?

1. Introduction

The intention of this book was outlined in the Preface. Some justification, therefore, is needed for devoting a chapter exclusively to regulation of plant enzymes. It is an entirely empirical decision. On the one hand, most reviews assume that only animal biochemistry is of interest (1). It is an exaggeration to say that calcium-based regulation and reversible phosphorylation are non-existent in plants, but certainly they are relatively much less important than in animals, and are also only just beginning to be intensively investigated. On the other hand, regulation by oxidation/reduction is very important in plant metabolism, especially in photosynthesis, while in animal cells it has never turned out to have the importance predicted for it when glutathione was first investigated. Thus the puzzling features of oxidation/reduction (redox) regulation can hardly be understood without reference to plant metabolism, especially as the redox status is so closely related to photosynthesis, and therefore to light/dark adaptation in plants. For comprehensive information about CO_2 fixation in plants, see Section 5.

Studies of control mechanisms have never been so popular among botanists as they are with biochemists studying metabolism in animals and bacteria, and therefore the details given in this chapter are more scanty than one would wish.

Figure 6.1 summarizes the essential features of chloroplast metabolism in 'C3' plants, in which labelled CO_2 first appears in phosphoglycerate (PGA). In 'C4' cells, it can first be detected in malate or oxaloacetate (OAA), and only later reappears in the reductive pentose phosphate ('Calvin') cycle. 'C3' chloroplasts contain $NADP^+$-linked malate dehydrogenase (MDH). In many C4 plants, this enzyme is found in the chloroplasts of specialized auxiliary cells (mesophyll cells), where it reduces the primary CO_2-containing product OAA to malate. This is

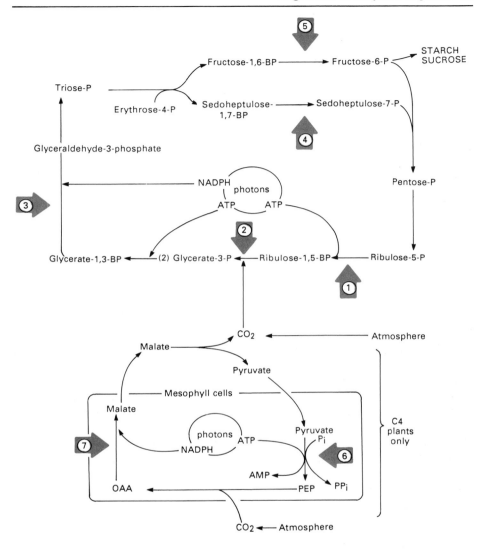

Figure 6.1. Simplified diagram of the Calvin cycle of the path of carbon in photosynthesis (upper part of the figure), and of the CO_2-trapping reactions in mesophyll cells, which only operate in C4 plants. The orange arrows indicate enzymes in chloroplasts which are sensitive to light/dark regulation. The numbers refer to *Table 6.1*.

then exported to the 'bundle sheath' cells, whose chloroplasts contain the Calvin cycle $NADP^+$-linked MDH and pyruvate phosphate dikinase (which occurs only in C4 plants and in bacteria) are two of the enzymes whose activation by light has been most intensively studied (2,3).

Although plants have the specialized enzymes of carbohydrate synthesis shown in *Figure 6.1*, they also have glycolysis (to pyruvate) and the oxidative pentose

Table 6.1. Chloroplast enzymes regulated by light/dark modulation

	Effect of light mediated by thioredoxin	Modulating metabolite (see *Figure 6.3*)
Reductive pentose phosphate pathway		
1. Phosphoribulose kinase	+	ATP ($-$)
2. Ribulose diphosphate carboxylase	[not affected by redox chain]	Carboxy-arabinitol-P and other metabolites ($-$)
3. NADP$^+$-linked glyceraldehyde-3-phosphate dehydrogenase (GAPDH)	+	1,3-diPGA; P_i (+)
4. Sedoheptulose bisphosphatase	+	Sedoheptulose-P_2 (+)
5. Fructose-1,6-bisphosphatase (FBPase)	+	FBP (+)
C4 pathway		
6. Pyruvate phosphate dikinase	+	ADP, PP_i ($-$)
7. NADP-linked MDH	+	NADP$^+$ ($-$)
Other pathways		
ATPase	+	Reduced glutathione
Phosphofructokinase	$-$	Reduced glutathione
Glucose-6-phosphate dehydrogenase		NADPH (+)

phosphate pathway found in animal cells. The problems associated with the initiation of glycolysis in plants are discussed in Section 3. Pyruvate oxidation, which is quite vigorous in plant mitochondria, is carried out by a pyruvate dehydrogenase complex which is regulated by reversible phosphorylation, just as in animal tissues, although the factors affecting the activation and inactivation of the complex in plants have not been intensively studied.

2. Regulation of enzymes by light

None of the enzymes that are about to be discussed are directly responsive to light; nevertheless, their activity changes, either completely or by at least 85%, within 3–5 min of a photosynthesizing tissue being placed in light or dark. Usually the change is in the direction of activation in light and, although the evidence is scanty, it does seem likely that the extent of activation is correlated with light intensity (4).

Not all enzymes whose activities are known to be regulated by changes in illumination are related directly to photosynthesis (5), as *Table 6.1* shows. However, many of the most important examples are found within chloroplasts. Since these organelles are specialized for the production of reducing power and ATP to drive the fixation of CO_2, it might be thought that most of the regulation is carried out by variations on a single mechanism, but this is not at all the case.

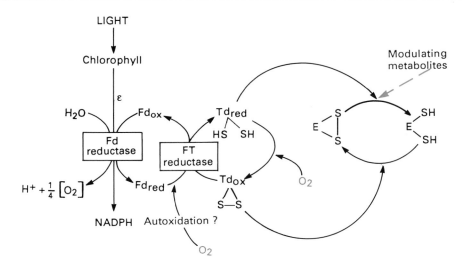

Figure 6.2. Mechanism of response of chloroplast enzymes to light. Fd is ferredoxin, an iron-containing protein in the photosynthetic electron transport pathway; Td is thioredoxin. Fd reductase, ferredoxin reductase; FT reductase, ferredoxin – thioredoxin reductase. E, MDH; S, thiol groups.

2.1 Thioredoxin-mediated regulation

Thioredoxins are small proteins (mol. wt 11 000) which contain one or more easily oxidizable and reducible cysteine groups. They do not contain iron, which distinguishes them from the ferrodoxins. A review on this topic has been published (6). Ribonucleoside diphosphate (rNDP) reductase (see Section 2 of Chapter 3) uses a thioredoxin as a primary reductant, and is itself strictly regulated, but in the mechanisms to be described, the thioredoxin is a regulating agent. It is thought that at least five of the enzymes shown in *Figure 6.1* and *Table 6.1*, NADP$^+$-linked glyceraldehyde-3-phosphate dehydrogenase (GAPDH), ribulose-5-phosphate kinase, sedoheptulose bisphosphatase, fructose bisphosphatase [but not the alternative pyrophosphate fructose-6-phosphate (F-6-P) phosphotransferase, see Section 3] and NADP$^+$-linked MDH, are activated by reduced thioredoxin$_m$ (Thio$_m$) or thioredoxin$_f$ (7). The relation between illumination and redox control of the enzymes is summarized in *Figure 6.2*.

NADP$^+$-linked MDH has two vicinal thiols near but not in the active site, which disappear on activation. Conversion to the disulphide forms causes a dramatic reduction in the affinity of the enzyme for NADP$^+$ and NADPH, but there is no incontrovertible information about conformational changes. *Figure 6.3* is based on sequence data, but is to some extent speculative.

A feature of the regulation apparently peculiar to NADP$^+$-linked MDH is that NADP$^+$ inhibits both activation and inactivation, while NADPH reverses this inhibition (2). The relative dissociation constants suggest that NADP$^+$ will quite strongly inhibit activation, while NADPH probably helps to prevent inactivation.

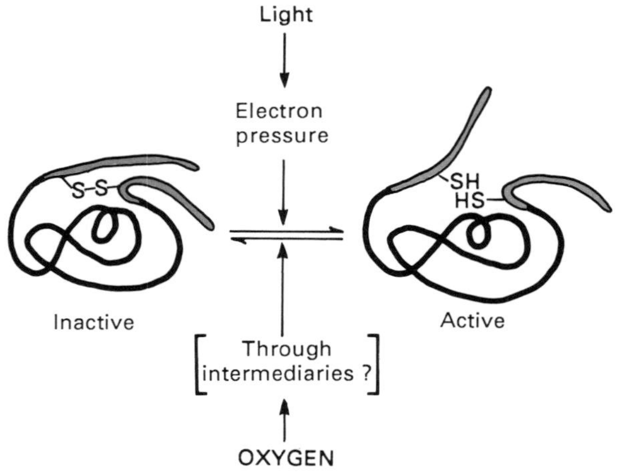

Figure 6.3.   Proposed interconversion of active and inactive forms of chloroplast NADP$^+$-dependent MDH. The ends of the protein chain coloured in orange are found only in the chloroplast enzyme (from ref. 3).

Thus the NADP$^+$:NADPH ratio can affect the enzyme in two ways, both directly, and indirectly by controlling the Thio$_m$(ox):Thio$_m$(red) ratio. Both mammalian and plant (see *Table 6.1*) glucose-6-phosphate (G-6-P) dehydrogenases, which are almost inactive in the absence of reduced glutathione, are also prevented from inactivation by NADPH (2), but NADP$^+$ plays no role.

Figure 6.3 leaves open to question the mechanism of inactivation of this and other thioredoxin-mediated enzymes in the dark. In C3 plants it is clear that NADP$^+$-dependent MDH is fully activated at quite low light intensities, and is dynamically protected from inactivation through oxidation (O$_2$ being the oxidant) by the electron pressure generated by the photosynthetic electron transport system. The *effective* capacity of this and other Calvin cycle enzymes, and their integration—the partition of metabolic control, to use the nomenclature of Chapter 2—is then superimposed on this by allosteric ligand binding of chloroplast metabolites. A good review of this complex field has been published (3). In C4 plants the picture is not so clear, because mesophyll cell chloroplasts do not contain the Calvin cycle intermediates, although they certainly are supplied with light-generated NADPH and ATP. Moreover, the range of light intensities over which full activation is reached is much greater: 3–50% full illumination, and there is also a strong temperature dependence of activation (4).

2.2 Mediation by reversible phosphorylation

2.2.1 Pyruvate phosphate dikinase

This enzyme catalyses the reaction between pyruvate (pyr), phosphate (P$_i$) and ATP thus:

$$\text{Pyr} + \text{P}_i + \text{AMP-P-P}^* \rightleftharpoons \text{PEP} + \text{P}^*\text{P}_i + \text{AMP}$$
$$\text{(ATP)}$$

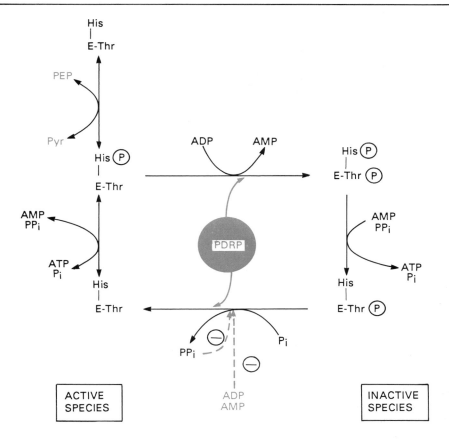

Figure 6.4.   Activation and inactivation of pyruvate, phosphate dikinase by pyruvate dikinase regulatory protein (PDRP). PDRP is a bifunctional enzyme, with two distinct active centres. The active centre of the dikinase contains a histidyl residue, which accepts a phosphoryl group from either ATP or PEP. Only the phosphorylated enzyme can be inactivated, while reactivation of this species is very slow, and is not shown in the diagram.

to form phosphoenolpyruvate (PEP), pyrophosphate (PP$_i$) and AMP.

The mechanism involves transfer of a phosphate group, P*, to a histidinyl residue in the active centre to form a phosphorylated enzyme intermediate.

Curiously, this enzyme also has two vicinal thiols near the active centre, but the mechanism of regulation is quite different from that discussed in the previous section, and has a unique feature; the enzyme is inactivated by phosphoryl transfer to a threonine residue near the active centre, but the donor is ADP, not ATP. The activation mechanism is also unusual, in that it involves phosphorolysis rather than hydrolysis, as shown in *Figure 6.4* (but see the inactivation of glutamine synthase, *Figure 4.7*).

Neither the phosphorylation nor the dephosphorylation reaction is functionally reversible, and both reactions are catalysed by a single regulatory protein (PDRP) (8). It is not known whether the latter has two catalytic sites, or if its own activity is subject to regulation. Current views (2) of the regulatory mechanism consequently depend heavily on changes in concentration of intracellular metabolites, as with $NADP^+$-dependent MDH (above). A powerful controller could be the ratio of pyruvate to PEP, since the only form of the enzyme which can be inactivated is that in which the histidine residue at the activated site is not phosphorylated, as shown in *Figure 6.4*. This would be more likely when the pyruvate concentration is high. The concentration of ADP is probably also very important, both because it is the phosphoryl donor for the inactivation, and also because it inhibits, more strongly than any other effector so far discovered, PDRP acting as a re-activating enzyme. When light is switched off, so that photosynthesis stops, there will be a rise in the concentration of ADP, although actually the K_m for ADP in the inactivation process is very low, about 50 μM (9), and it does not seem likely that the concentration of ADP would ever fall so low that, by itself, it would bring the inactivation process to a halt.

It is fair to say that these mechanisms are so far speculative (8). Pyruvate accumulates in anaerobic leaves in the dark, as might be expected, but not simply in darkened leaves, because there is a strong mitochondrial respiration in the dark, for which pyruvate is known to be a good substrate. This respiration, moreover, produces ATP very efficiently, while the concentration of ATP, and therefore of ADP, depends not only on its rate of synthesis, but also on its rate of utilization. These problems could only be solved by measurements of metabolite changes in mesophyll cells themselves, not just in rapidly darkened leaves, and this is technically a rather difficult problem, as might be imagined. Also it would be very unusual, to judge from control systems in animals, if the regulating enzyme PDRP were not itself regulated, at least by an allosteric ligand, but no evidence for this is available.

This brief discussion shows how difficult it is to allocate control mechanisms with certainty in the absence of very detailed measurements of changes in metabolite concentrations. There are other enzymes in plants which are regulated by phosphorylation; a review has been published on this topic (10).

2.2.2 Quinate:NAD^+ oxidoreductase

This enzyme catalyses a step in a pathway of aromatic amino acid synthesis, which is of course important in plants. It is activated when it is phosphorylated by a calmodulin (CaM)-dependent protein kinase, and is indeed one of the few examples at present known of CaM-based regulation in plants (11). Moreover, in the dark a subunit that binds Ca^{2+} becomes attached to the catalytic subunit, and the enzyme is then directly regulated through $[Ca^{2+}]$. This double mechanism for ensuring regulation by $[Ca^{2+}]$ is reminiscent of, but not exactly the same as, phosphorylase kinase and protein phosphatase 2B (see Section 1.2 of Chapter 4).

2.3 Mediation by effector binding

There are two well-known examples: $NADP^+$-linked GAPDH and ribulose diphosphate carboxylase (familiarly known to botanists. as Rubisco); both are chloroplast enzymes (6).

The response of $NADP^+$-linked GAPDH to light is fairly straightforward; both ATP and NADPH, whose concentrations increase in light, are positive effectors. Inorganic phosphate brings the binding isotherm for both ligands into the physiological range of concentrations. These controls are in addition to those brought about by thioredoxin on this enzyme (*Table 6.1*).

It is more difficult to be certain about the regulation of Rubisco. As so many of the enzymes of the Calvin cycle are inactivated in the dark, there seems to be little need for Rubisco inactivation; it may be perhaps that it is the oxygenase function that is suppressed. At all events it is accepted that the enzyme is much less active in the dark than in the light; so far, ligand binding is the only known regulatory mechanism.

It has been shown that CO_2 will react reversibly with a lysine residue on the large subunit of the enzyme, which can bind Mg^{2+} to form an active enzyme – $CO_2 - Mg^{2+}$ complex. The formation of the carbamate would be favoured by the more alkaline pH and increased $[Mg^{2+}]$ in photosynthesizing chloroplasts. It is now considered unlikely that changes in pH and $[Mg^{2+}]$ *in vivo* would be large enough to have a marked effect on the regulation of chloroplast enzymes (6), and with Rubisco, changes in activity cannot be correlated with changes in concentration of these ligands.

In vitro, the enzyme is inhibited by a number of allosteric ligands including ATP, NADH, 6-phosphogluconate, and intermediates of the photosynthetic pentose phosphate cycle (6). Such a large list usually means that the major controlling mechanism has not been identified, and it may be that Rubisco is regulated by a covalent modification. However, a new sugar derivative has been identified (12) that is present only in darkened leaves, namely 2-carboxy-D-arabinitol-1-phosphate. In spite of its terrifying name, it is simply related to the immediate carboxylation product of ribulose bisphosphate, itself only recently proved to have an independent existence.

Compound C (see *Figure 6.5*) binds almost irreversibly to Rubisco much as if it were a transition state inhibitor, and the presumption is that compound D binds almost as strongly. However, for D to be effective it must bind in a one-to-one ratio with the enzyme; the concentration of Rubisco in C3 chloroplasts and bundle sheath cells is about 10^{-4} M, and the inhibitor would have to be present at a somewhat higher concentration to be effective. More to the point, although there is typically a lag period before CO_2 fixation is maximal after illumination, the inhibitor would have to disappear within a few minutes. How could such a relatively enormous quantity of metabolite be disposed of?

To pose this question is to stress once again that physiologically important regulation is not only a matter of finding a suitable ligand, but of demonstrating that it can be both formed and destroyed at a suitable rate in the relevant conditions *in vivo*. This is clearly more easily achieved by the initiation of a cascade (see Chapter 1, Section 1).

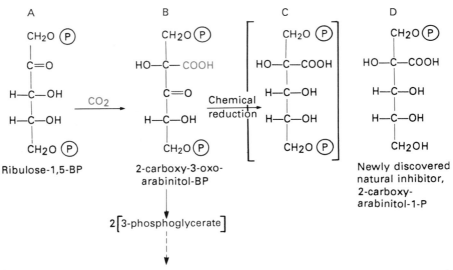

Figure 6.5. Carbohydrates that interact with ribulose diphosphate carboxylase. (**A**) is the substrate, while (**B**) is the immediate product of carboxylation, which has only just been demonstrated to exist, since it normally disproportionates very rapidly into two molecules of 3-phosphoglycerate. (**C**) can be formed by chemical reduction of B, and is an almost irreversible inhibitor of the enzyme. (**D**) is a natural product, which accumulates only in darkened leaves, and is a strong inhibitor of the enzyme.

3. Fructose-2,6-bisphosphate in plants

Fructose-2,6-bisphosphate, the importance of which in the regulation of carbohydrate metabolism in liver has been discussed in Section 2.5 of Chapter 5, is also found in plants (13–15), but only in the cytosol. The fructose bisphosphatase found in plant cytosol is sensitive to inhibition by F-2,6-BP, but this enzyme coexists with a higher concentration of pyrophosphate-F-6-P phosphotransferase (Section 2.5 of Chapter 5), which is found only in plants and in some bacteria. F-2,6-BP is just as powerful an activator of this enzyme as it is of animal PFK. In any event, the ATP-linked PFK, found only in chloroplasts, is insensitive to F-2,6-BP. Nothing could indicate more clearly that chloroplast metabolism is subject to its own special constraints, in which the presence or absence of light is much more important than energy demand or the type of nutrient available, and regulation has evolved accordingly.

The existence in high concentration of a reversible enzyme which is very sensitive to a regulating ligand is something of an embarrassment to botanists because the activation cannot be the controller (in the Newsholmian sense) of either glycolysis or gluconeogenesis. At the time of writing a number of hypotheses, none very convincing, are being tentatively put forward (16).

4. General comments

Even a cursory glance at plant metabolism must provide a salutary shock, particularly to cherished views about 'committed steps' and the importance of irreversible enzymes.

At the same time, CO_2 fixation does appear to be rather over-regulated, as *Figure 6.1* and *Table 6.1* indicate. One could very much like to see metabolic flux analysis applied to the pathways summarized in this diagram (*Figure 6.1*); one would, however, be lost in admiration for the biochemist who undertook an experimental investigation. The technical problems, even in isolated chloroplasts, would be formidable.

5. Further reading

Edwards,G.E. and Huber,S.C. (1981) In *The Biochemistry of Plants*, Volume 8. Hatch,M.D. and Boardman,N.K. (eds), Academic Press, New York, p. 237.
Stitt,M., Huber,S.C. and Kerr,P. (1988) In *The Biochemistry of Plants*, Volume 10. Hatch,M.D. and Boardman,N.K. (eds), Academic Press, New York, in press.

6. References

1. Wraight,C.A. and Cheeseman,J.M. (1987) *Trends Biochem. Res.,* **12**, 101.
2. Edwards,G.E., Nakamoto,H., Burnell,J.N. and Hatch,M.D. (1985) *Annu. Rev. Plant Physiol.,* **36**, 255.
3. Scheibe,R. (1987) *Physiol. Plant.,* **71**, 393.
4. Hatch,M.D. (1978) *Curr. Top. Cell. Regul.,* **14**, 1.
5. Anderson,L.E. (1986) *Adv. Bot. Res.,* **12**, 1.
6. Buchanan,B.B. (1980) *Annu. Rev. Plant Physiol.,* **31**, 341.
7. Holmgren,A. (1985) *Annu. Rev. Biochem.,* **54**, 237.
8. Burnell,J.N. and Hatch,M.D. (1985) *Arch. Biochem. Biophys.,* **237**, 490.
9. Buchanan,B.B. (1984) *Bioscience,* **34**, 378.
10. Kauss,H. (1987) *Annu. Rev. Plant Physiol.,* **38**, 47.
11. Ranjeva,R. and Boudet,M. (1987) *Annu. Rev. Plant Physiol.,* **38**, 73.
12. Foyer,C. (1986) *Nature,* **324**, 274.
13. Van Schaftingen,E. (1987) *Adv. Enzymol.,* **59**, 315.
14. Huber,S.C. (1986) *Annu. Rev. Plant Physiol.,* **37**, 233.
15. Stitt,M. (1987) *Plant Physiol.,* **84**, 201.
16. Fritz,G., Vogelmann,T.C. and Bornman,J.F. (eds) (1987) Symposium Report in *Physiol. Plant.,* **69**, 373.

7

Overview

'I do not care to pronounce an opinion on this matter; I leave it as an
interesting speculation'
Norbert Wiener

1. What happens inside cells?

One of the most difficult problems to face, in writing a book of this kind, is the
argument that theorems of regulatory control elaborated *in vitro* are irrelevant
to intracellular metabolism, because enzymes are packed so tightly in cells that
they no longer behave as individual entities, but as transient or permanent
supermolecular clusters, to which the rules of enzyme behaviour elaborated *in
vitro* need not exactly apply. Some reference to this attitude has been made in
Chapter 3. Since it is almost impossible to study this 'microcompartmentation'
in vitro, the argument, if pushed to its logical conclusion, amounts to the re-
introduction of a 'vital force', an idea prevalent right up to the beginning of the
20th century, and one which would make the study of cell biology at the molecular
level rather pointless. Only enzyme kinetics would remain a valid study, because
many enzymes are extracellular, and because isolated purified enzymes can be
used as industrial chemicals.

The argument about 'microcompartmentation' cannot be dismissed out of hand.
In order to come to a reasoned view, I propose to approach the subject rather
circuitously.

1.1 Bifunctional proteins

Five regulated or regulatory proteins are known at present which are bifunctional,
presumably as a result of gene fusion; they are listed in *Table 7.1*. Several other
bifunctional enzymes are known which have not been included in the table
because they do not form part of a regulatory cascade. A well-studied example
is bacterial aspartate kinase I/serine dehydrogenase I, which is regulated by end-

Table 7.1 Bifunctional regulated or regulatory proteins

Protein	Regulated by
1. Bacterial isocitrate dehydrogenase kinase/phosphatase (Chapter 2)	metabolite ligands
2. Fructose-6-phosphate (2)-kinase/phosphatase (Chapter 5)	cAMP (liver, yeast), metabolite ligands
3. Adenyl transferase (AT) to glutamine synthase (Chapter 4)	metabolite ligands
4. Uridyl transferase to AT (Chapter 4)	metabolite ligands
5. Pyruvate phosphate dikinase regulatory protein (Chapter 6)	none known

product feedback by the amino acids whose synthesis it initiates (1). The proteins listed in *Table 7.1*, and their modes of regulation, have little in common with one another. It might be thought that a ligand which allosterically activates one site and inactivates another on the same peptide chain, would provide a flip-flop effect, but computer simulations by Stadtman's group (2) do not confirm this. They do suggest that a closed bicyclic cascade, which is the inevitable consequence of a bifunctional regulatory protein, has far greater flexibility of response to metabolite modulation, and greater signal amplification, than an open cascade (see Chapter 1, Section 2.1).

1.2 Multifunctional proteins

Several multifunctional proteins are known (3). Two will be discussed here; the *arom* complex of *Neurospora*, *Saccharomyces* and *Aspergillus* (4), which catalyses a sequence of five of the seven steps of the shikimate pathway of aromatic amino acid synthesis, and the eukaryotic fatty acid synthase complex (5). In both instances, the same metabolic sequences are carried out in other organisms (e.g. *Escherichia coli*) by separate enzymes.

Several advantages are usually quoted for multifunctional enzyme organization:
(i)   enhancement of catalytic activity (catalytic facilitation);
(ii) substrate channelling;
(iii) coordinate regulation of enzyme activity;
(iv) protection of unstable intermediates;
(v)   coordinated expression of genes;
(vi) essential spatial organization of the enzymatic functions.

The *arom* complex is particularly interesting in the light of these postulates, because only proposition (v) is certainly valid (4). Moreover, the advantage of coordinated gene expression is only evident because the turnover numbers of the five enzymes all lie within a very narrow range. This is a point often forgotten by those who propound the advantages of disorganized enzyme clusters (see Section 1.3); there is little point in spatial organization if the turnover numbers of the enzymes vary widely. In *organized* multienzyme complexes, such as the pyruvate dehydrogenase complex, the pre-arranged differences in the numbers of catalytic sites no doubt take care of disparities in turnover number.

In the yeast fatty acid synthase complex (5), there are two large peptides, one of which (β) contains five enzyme activities; the other (α) two and the acyl carrier protein. Postulates (ii) and (vi) apply to these two peptides, and also (iv), although in a modified form which may be put as follows. Coenzyme A (CoA) is present in a fixed and limited quantity in cells and is distributed among a large number of thioesters (6). It is easy to see that a build-up of a particular fatty acyl thioester could produce a bottle-neck in a completely different pathway. In fatty acid synthesis this is avoided by transfer of the growing acyl group to the acyl carrier protein, which provides both spatial geometry (the 'swinging arm') and a coordinated—and locally high—concentration of the very reactive intermediates, together with protection for them. It has, moreover, been proposed (7) that the activity of the complex is regulated by partial removal of the pantotheine moiety of the acyl carrier protein.

However, in bacteria, fatty acid synthesis is catalysed by a series of separable, completely soluble, enzymes (and a carrier protein), and there is as yet no evidence that an enormous advantage is conferred by the multifunctional protein complex of eukaryotes (7).

1.3 'Disorganized' multienzyme complexes

This is perhaps an unfortunate description, but it means complexes that may exist intracellularly with no fixed ratio between the numbers of the different protein molecules, and which cannot be extracted predictably as entities in the same way as the oxoacid dehydrogenase complexes. The 'glycogen particle' of striated muscle (8) is one good example; others in this area of metabolism are the histologically demonstrable association of some glycolytic enzymes with myofibrils, and the association of some enzymes of triose phosphate metabolism with the red cell membrane. All of these have been reviewed in ref. 9, which concluded, after an exhaustive survey of the evidence, that only in the 'small' glycogen particles of muscle, containing only a small proportion of the enzymes of glycogenolysis, is there evidence for a functional advantage in the association of the enzymes.

The glycolytic particle of *Trypanosoma brucei* (10) is a particularly interesting example, because the enzymes are actually enclosed within a fragile membrane. On close examination, the functional value of this sac appears dubious. The enzymes from enolase onwards are completely missing, so that the assemblage can at best only make as much ATP [from the oxidation of glyceraldehyde phosphate (GAP)] as it consumes (in the phosphorylation of glucose and fructose-6-phosphate), and then only if some acceptor other than GAP is available for the NADH that is formed. In addition, it is not clear that 3- phosphoglycerate or 2-phosphoglycerate (which would normally be the substrate for enolase) can be rapidly exported through the membrane of the sac. Failing this, they can only accumulate inside. Many of the enzymes are in any case distributed evenly between the sac and other cellular compartments (10). In view of the wide disparity in turnover numbers, and the evidence that there is no channelling (proposition ii) at normal glucose concentrations (11), one must remain sceptical about the biological significance of this entity.

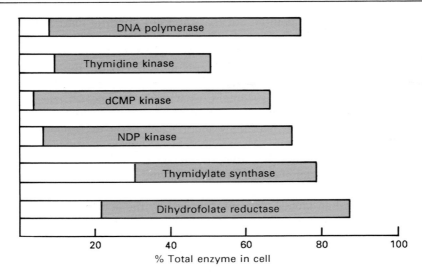

Figure 7.1. Percentage of enzyme activities found in the nuclear region of Chinese hamster embryo fibroblast cells. All the enzymes shown, except DNA polymerase, are concerned with the synthesis of deoxypyrimidine nucleotides, but later work has shown that NDP reductase is also found in such a complex, for which the name 'replitase' has been proposed (see text). The clear bars show the fraction of the enzymes sedimenting down with the nuclear region in quiescent cells, while the orange bars indicate the totals in growing cells. (Redrawn from ref. 13.)

The 'replitase' of bacteria and cultured animal cells (12) is the most convincing example of a disorganized association of enzymes with clear evidence of functional importance. Although some authors (13) are not convinced of the existence of a complex, there is very good evidence that some cytosolic nucleoside triphosphates, especially GTP, are not in equilibrium with the immediate precursors of newly forming DNA (12,14). It is reasonable to conclude that reduction of nucleoside diphosphates (NDPs) by NDP reductase is sequestered from the cytosol or nuclear fluid, so that newly formed deoxynucleoside triphosphate (dNTP) molecules are concentrated around DNA polymerase at the replication fork, a clear functional advantage. In certain cells a 'replitase complex' can be isolated during growth, but not in the resting phase. *Figure 7.1* shows the concentrations of the six enzymes which can be found in the complex when it is centrifuged down, as a percentage of the total concentration of each enzyme in the cells. There is also evidence that NDP reductase is also present in the complex when it forms.

From the regulatory point of view, it is very interesting that the h- and l-sites of NDP reductase can be affected by the same feedback metabolites, chiefly dATP (12), that are important *in vitro* (*Figure 3.1*). This implies both that the enzyme is subject to the same controls *in vivo* as *in vitro*, and that the replitase is organized to the extent that the regulatory sites, but not the catalytic sites, of the enzyme are exposed to the intracellular environment.

It has frequently been pointed out that the mitochondrial matrix contains a very high concentration of enzymes; thus *a priori* the attraction of proposing some kind of organization is considerable. The evidence for the functional effectiveness of close packing is surprisingly meagre (15,16); the most striking findings are more rapid attainment of a new steady state, and enhanced affinity of succinyl CoA synthase for succinyl-CoA which might nevertheless imply a reduction in catalytic efficiency (see Section 3 of Chapter 3). One must also take into account that in ureotelic liver mitochondria 20% of the mitochondrial matrix consists of the single enzyme carbamoylphosphate synthase, the concentration of which is known to change with the protein content of the diet (17); this must markedly affect the packing density. The enzyme has complex kinetics, but metabolic flux analysis shows that urea cycle flux is entirely controlled by the intramitochondrial ammonia concentration and by the activity of the synthase (17). Studies with permeabilized liver mitochondria suggest that the kinetic properties of the synthase *in situ*, when substrate and enzyme concentrations are about equal (≈ 0.4 mM), are very similar to those in dilute solution (18). This latter approach—to see whether metabolic rates, both with and without regulation, agree with those measured *in vitro*—has chiefly been used with bacteria. In the meticulous experiments of Koshland (19) on citrate oxidation in *E.coli* (see also Chapter 4), and the studies of glutamine synthase regulation in permeabilized *E.coli* (20), the agreement between theory and experimental measurement is very good indeed.

It is, in any case, not automatically true that the closer the packing of the enzymes, the more rapid the transfer of intermediates between them. If transfer takes place by diffusion, very close packing can decrease transfer rates. This is because, although the diffusion distance is reduced, for a fixed quantity of enzymes the *gradient* necessary to maintain a given flux is increased, and may not be maintainable in three-dimensional close packing (21). The increased gradient can lead to product inhibition of a prior enzyme, and substrate limitation in the following one.

To summarize, the evidence, although admittedly patchy, suggests that it would be unnecessarily alarmist to think that 'microcompartmentation', as a general rule, vitiates the application either of metabolic flux analysis or of knowledge about the regulation of individual enzymes obtained from *in vitro* studies. There may be exceptions to this judgement in animal metabolism, for instance 'replitase', and also mitochondrial electron transport (22). However, other membrane-bound enzymes such as hydroxymethylglutaryl-CoA (HMG-CoA) reductase behave predictably with respect to allosteric inhibitors and reversible phosphorylation (23).

2. Panglossian biology

'All is for the best in the best of all possible worlds' *Voltaire*

In his masterpiece *Candide*, Voltaire created a Dr Pangloss, whose favourite saying is quoted above, to poke fun at the perfectionist philosophers of his time.

We are all inclined to take a Panglossian attitude towards biological phenomena. How far is this justifiable? One of the most striking things about regulated enzymes is that their activity is very often controlled by more than one mechanism. Indeed, one might create a rule-of-thumb to the effect that every really important flux-controlling enzyme has at least two control mechanisms. HMG-CoA reductase, for instance, is not only subject to feedback control by cholesterol and allosteric regulation by other steroid metabolites, but is also regulated by a reductase kinase and phosphatase; in addition, the enzyme has a short half-life with a marked diurnal rhythm of synthesis. This example is not unique; are all the controls on every enzyme always of great importance?

The 'belt and braces' approach may, at least partly, be due to the fact that we are looking at a 'fossil record' of control evolution. One of the motives in writing a book of broad scope, with unusual emphasis on plant and bacterial enzymes, was to draw attention to the very wide variety of ways in which modulation may be used to achieve a single end. Actomyosin contraction is a good example; it is initiated, in different cells or organisms, by three different mechanisms, all involving Ca^{2+}. In the 'catch' muscle of molluscs, the calcium interacts directly with myosin light chains; in skeletal muscle, it interacts through troponin C; and in smooth muscle and platelets, it activates, through calmodulin, myosin light chain kinase, which initiates contraction through phosphorylation (see Section 1.1.1 of Chapter 4). Cyclic AMP, in contrast, has quite different roles in eukaryotes and prokaryotes. If, as a result of evolutionary pressure, modulation of any enzyme becomes more appropriate with a different effector, the likelihood surely is that, unless it is positively harmful in the new circumstances, the original regulatory domain will remain. Anyone who has ever had to deal with a computer program of even moderate size will know that it contains labels and routines with no possible further usefulness, but only a hero would attempt to edit them out. The attempt to introduce aesthetically pleasing tidiness would almost certainly introduce new 'bugs': better leave well alone. The metaphor is not inapt.

Certainly in endocrinology it is true that hormones often interact inefficiently, for example in regulation of salt and water balance by mammalian kidney. It is probable that some of the controls are survivors, left over from quite different circumstances, for example the regulation of salt and water efflux through amphibian skin. To take another example, insulin is manufactured and secreted in animals in which regulation of blood sugar is only of marginal importance, for example fish, poultry and sheep.

The obverse of this viewpoint is that there always seem to be more enzymes whose activity is capable of being regulated, in most areas of metabolism, than can possibly be needed for the purpose. Here again is a fossil record. We must not assume that everything that we now observe in a regulated metabolic system developed simultaneously. Metabolic pathways themselves adapt to changed environments. A glance back at the various enzymes that catalyse the synthesis of fructose-1,6-bisphosphate (Chapter 5) should make this quite clear.

In the end, however, with such a plethora of facts from which to choose, people select, and believe what they want to believe, which is usually a logical, tidy, aesthetically acceptable and intellectually comforting simplification. Cajals, one of the early Nobel prize-winners, put it impressively in his autobiography: 'I wish to warn young [persons] against the invincible attraction of theories which simplify and unify seductively . . . We fall into the trap all the more readily when the simple schemes stimulate and appeal to tendencies deeply rooted in our minds, the congenital inclination to economy of mental effort and the almost irresistable propensity to regard as true what satisfies our aesthetic sensibility by appearing in agreeable and harmonious architectural forms. As always, reason is silent before beauty' (24).

Not much can be done about this weakness of ours, but the message of this book is that it is often possible, and always desirable, to measure quantitatively the extent to which our simple schemes match with the reality of the observable world.

3. Further reading

Welch,G.R. (ed.) (1985) *Catalytic Facilitation in Organized Multienzyme Systems.* Academic Press, New York.

4. References

1. Truffa-Bachi,P., Veron,M. and Cohen,G.N. (1974) *CRC Crit. Rev. Biochem.,* **2**, 379.
2. Stadtman,E.R. and Chock,P.B. (1978) *Curr. Top. Cell. Regul.,* **13**, 53.
3. Kirschner,K. and Bisswanger,H. (1976) *Annu. Rev. Biochem.,* **45**, 143.
4. Coggins,J.R., Duncan,K., Anton,I.A., Boocock,M.R., Chaudhuri,S., Lambert,J.M., Lewendon,A., Millar,G., Mousdale,D.D.M. and Smith,D.D.S. (1987) *Biochem. Soc. Trans.,* **15**, 754.
5. Singh,N. and Stoops,J.K. (1987) In *Enzyme Mechanisms.* Page,M.I. and Williams,A. (eds), R. Chem. Soc., London, p. 534.
6. Hammes,G.G. (1985) *Curr. Top. Cell. Regul.,* **26**, 311.
7. Bloch,K. and Vance,D. (1977) *Annu. Rev. Biochem.,* **46**, 263.
8. Taylor,C., Cox,A.J., Kernohan,J.C. and Cohen,P. (1975) *Eur. J. Biochem.,* **51**, 105.
9. Ottaway,J.H. and Mowbray,J. (1977) *Curr. Top. Cell. Regul.,* **12**, 107.
10. Opperdoes,F.R. and Borst,P. (1977) *FEBS Lett.,* **80**, 360.
11. Aman,R.A. and Wang,C.C. (1986) *Mol. Biochem. Parasitol.,* **19**, 1.
12. Pegg,A.E. (1986) *Biochem. J.,* **234**, 249.
13. Reichard,P. and Nicander,B. (1984) *Curr. Top. Cell. Regul.,* **26**, 403.
14. veer Reddy,G.P. and Pardee,A.P. (1980) *Proc. Natl. Acad. Sci. USA,* **77**, 3312.
15. Srere,P.A. (1980) *Trends Biochem. Res.,* **5**, 120.
16. Srere,P.A. (1987) *Annu. Rev. Biochem.,* **56**, 89.
17. Meijer,A.J. (1985) In *Metabolic Regulation.* Ochs,R.S., Hanson,R.W. and Hall,J. (eds), Elsevier, Amsterdam, p. 171.
18. Lof,C., Cohen,M., Vermeulen,L.P., Van Roermund,C.W.T., Wanders,R.J.A. and Meijer,A.J. (1983) *Eur. J. Biochem.,* **135**, 251.
19. Koshland,D.E., Walsh,K. and LaPorte,D.C. (1985) *Curr. Top. Cell. Regul.,* **27**, 13.
20. Mura,U., Camici,M. and Gini,S. (1985) *Curr. Top. Cell. Regul.,* **27**, 233.
21. Ottaway,J.H. (1983) *Biochem. Soc. Trans.,* **11**, 47.
22. Westerhoff,H.V., Groen,A.K. and Wanders,R.J.A. (1984) *Biosci. Rep.,* **4**, 1.
23. Gibson,D.M. and Parker,R.A. (1987) In *The Enzymes,* Volume 18B. Boyer,P.D. and Krebs,E.G. (eds), Academic Press, New York, 3rd edn, p. 180.
24. Quoted by Huxley,A.F. (1977) In *The Pursuit of Nature.* Cambridge University Press, Cambridge, p. 23.

Glossary

Active centre: the site(s) within the tertiary or quarternary structure of an enzyme where binding of substrates and subsequent reaction occurs.

Agonist: a physiological effector or drug which binds to a receptor and initiates a biological process.

Allosteric ligand: a molecule which binds to a site on an enzyme, distinct from the active centre, causing a conformational change which alters the kinetic properties of the enzyme.

Cyclic cascade: an enzyme which can be reversibly activated and inactivated by two converter enzymes acting in opposition. The latter may also be reversibly activated and inactivated to form a polycyclic cascade.

Feedback: a control device in which information about a product is transmitted *upstream*, against the flux of energy or metabolites, to a regulatable producer step.

Feedforward: a regulatory device in which the control information is transmitted *downstream*.

Flux: (symbol J) the flow of energy or metabolites along a recognizable pathway, typically, but not always, in a steady state.

Flux control coefficient: the sensitivity coefficient (q.v.) of an enzyme or transporter in a defined segment of a metabolic pathway in a steady state.

Half reaction time: (symbol $t_{1/2}$) the time taken for a non-reversible first-order reaction to be half completed (for radioactive decay, the half-life).

Heterotropic cooperativity: see homotropic cooperativity.

Homotropic cooperativity: a change in conformation of subunits in a multi-subunit protein in which all the ligands that induce the change (typically substrate molecules) are identical. In *heterotropic cooperativity* the ligands are different (e.g. substrate and allosteric effector).

Hyperbolic Kinetics: the reaction kinetics typical of enzymes which obey the Michaelis–Menten equation, which (in the absence of products) describes a rectangular hyperbola.

K_m: the Michaelis constant for a single-substrate enzyme with hyperbolic kinetics. For an enzyme with cooperative kinetics, K_m has to be replaced by $K_{50\%}$, the substrate concentration at which 50% of the limiting rate is obtained (see V).

Induced fit model: the theory, proposed by Koshland, that the active site of an enzyme changes conformation only *after* the substrate binds, in such a way as to reduce the activation energy of the reaction.

'Magic spots': magic spots I and II were first identified by thin-layer chromatography. They are guanosine 5′ diphosphate 3′ diphosphate and guanosine 5′ triphosphate 3′ diphosphate, respectively. They accumulate in bacterial cells during amino acid starvation and mediate the stringent response.

Metabolic control analysis: the analysis of the distribution of control among the enzymes in a metabolic pathway, in a steady state under defined conditions, by measuring their flux control coefficients (q.v.) and applying the summation equations (see Chapter 2).

Respiratory control: the dependence of the rate of oxygen uptake by isolated mitochondria on ADP concentration; it can be given a quantitative value from experimental observations.

Sensitivity coefficient: the fractional change in flux through an enzyme produced by a fractional change in the concentration of that enzyme.

Signal amplification: the concentration of modulator ligand required to convert 50% of a converter enzyme to its active form, divided by the concentration of ligand required to convert the interconvertible enzyme I to its modified form.

State 3: State 4: these classify some of the respiratory states of mitochondria as follows:

state 1: no substrate, no ADP, respiration low.

state 2: no substrate, ADP, respiration low.

state 3: substrate, ADP, respiration high.

state 4: substrate, ADP exhausted, respiration low.

Index